# Overpressured Gas Reservoirs

**Steven W. Poston**
*Professor Emeritus, Texas A&M U.*

and

**Robert R. Berg**
*Michel T. Halbouty Chair, Retired, Texas A&M U.*



**Society of Petroleum Engineers, Inc.**
**Richardson, Texas, U.S.A.**
**1997**

# SPE Books Committee





**ISBN 1-55563-075-8**

# Preface

Overpressured reservoirs are abundant in sedimentary basins throughout the world. These reservoirs commonly produce light oils and gases and require special evaluation techniques. The purpose of this book is to describe the major characteristics of these reservoir from both the geologic and engineering viewpoints. Chap. 1 describes the geologic conditions for the origins of overpressured reservoirs. Knowledge of the origins is helpful in detecting and predicting the overpressured condition. Chap. 2 describes the effects of overpressuring on such reservoir parameters as porosity, permeability, and compressibility and on other properties important to hydrocarbon production.

Chaps. 3 through 6 develop the engineering aspects of studying overpressured reservoirs and suggest methods of evaluation. Many of these reservoirs are fully bounded in thick sections of shale; consequently, pressure-depletion conditions should exist. In actuality, a study of the production histories of overpressured gas reservoirs indicates that water influx of varying degrees of effectiveness is a fairly common occurrence. The material-balance equation has been modified to include water-influx effects that can profoundly affect the ultimate recovery estimate.

Methods have also been developed to replace static pressure normally required to develop pressure/fluid-property relationships in the material-balance equation with short-term measured pressures. This concept converts the material-balance method to a more economically viable evaluation tool. Calculations of original gas in place can be made from the early production history by means of the Becerra-Arteaga method, which is not widely used at the present time.

We thank the many geologists, engineers and students who have participated in discussing and testing the ideas and methods proposed in this book. We hope that this summary of overpressured reservoirs be a useful tool in developing oil and gas reserves throughout the world.

S.W. Poston
R.R. Berg

# Contents

# Chapter 1
# Occurrence and Origins of Overpressures

Overpressures are subsurface fluid pressures that are greater than the pressures normally expected under normal hydrostatic conditions. Prior knowledge of the possibility of encountering overpressures at particular subsurface depths is important when exploring for oil and gas. The presence of higher than normal pressure increases the complexity and cost of drilling, well-completion, and production operations. Additionally, the effect of overpressures on reservoir behavior must be recognized when predicting performance.

This chapter presents a summary of the geologic occurrence and possible origins of overpressures to help the reader understand the character of overpressured reservoirs. A discussion of the functional relationship between net overburden pressure and the combined influences of the effective stress exerted by the rock matrix and the pore pressure provides a background for Chap. 2.

## 1.1 Pressure Gradients

Subsurface fluid pressure profiles characterize the presence or absence of overpressured conditions. Measured pressures are commonly expressed in terms of the weight per unit area and represent the rise to the surface of a column of borehole fluid from the point of measurement in the subsurface. For example, a column of fresh water with 1.0-g/cm$^3$ density would exert a pressure gradient of 0.433 psi/ft, or 4,330 psi at a depth of 10,000 ft. A column of saline water of 1.07-g/cm$^3$ density would exert a pressure gradient of $(0.433)(1.07) = 0.465$ psi/ft, which translates to a 4,650 psi pressure at a depth of 10,000 ft. These pressures are called normal hydrostatic pressures if the water throughout the section is either essentially fresh or saline water. In the same way, a borehole filled with oil of average density of 0.8 g/cm$^3$ would have a hydrostatic pressure gradient of 0.346 psi/ft. A borehole filled with gas of density of 0.1 g/cm$^3$ would exhibit a pressure gradient of 0.043 psi/ft.

Designating pressure gradients as measured for a column of fluid in a borehole is a convenient way to express pressure measurements. In this case, the pressure gradients do not reflect the change in pressure over a specified depth interval but are an average, originating at the point of measurement in the subsurface and extending to the surface. The pressure gradients described in the following always refer to borehole gradients.

The pressure exerted by the total column of rock and fluids at a given depth is called the lithostatic pressure. These pressures cannot be measured conveniently in the subsurface but are easily estimated. For example, the average density of water-saturated sedimentary rocks in the U.S. gulf coast basin is about 2.3 g/cm$^3$. Therefore, the lithostatic

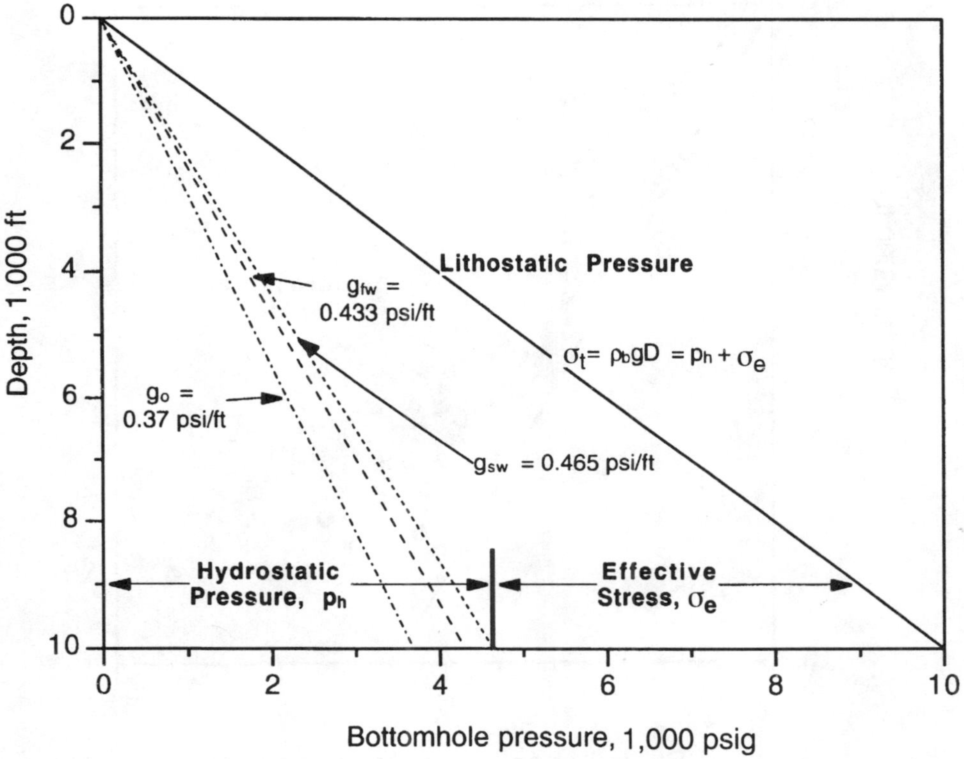


**Fig. 1.1—Definition of borehole pressure gradients for different materials.**

pressure gradient is (0.433 psi/ft)(2.3 g/cm$^3$) = 1.0 psi/ft of depth. Densities of the total sedimentary section are not uniform but increase with depth of burial from very low values to greater than 2.6 g/cm$^3$. Nevertheless, a lithostatic gradient of 1.0 psi/ft is a satisfactory approximation for the normal sedimentary section. **Fig. 1.1** illustrates the relation between the lithostatic stress and different single-fluid hydrostatic gradients. Normal hydrostatic-pressure gradients for salt water, fresh water, and a 34°API oil are compared. Overpressured reservoirs occur in the region lying between the lithostatic and the water hydrostatic gradient boundary lines. The lithostatic or overburden stress, $\sigma_t$, is the sum of the pore pressure and supporting strength of the rock matrix.

**Fig. 1.2** shows the effect of encountering overpressured zones along the Texas and Louisiana gulf coast. The saltwater and lithostatic gradients form the boundary lines. There is a pronounced gradient buildup to increasingly overpressured conditions at depths < 10,000 ft. The data spread reflects the wide range of possible conditions that may be encountered along the U.S. Gulf of Mexico coast.

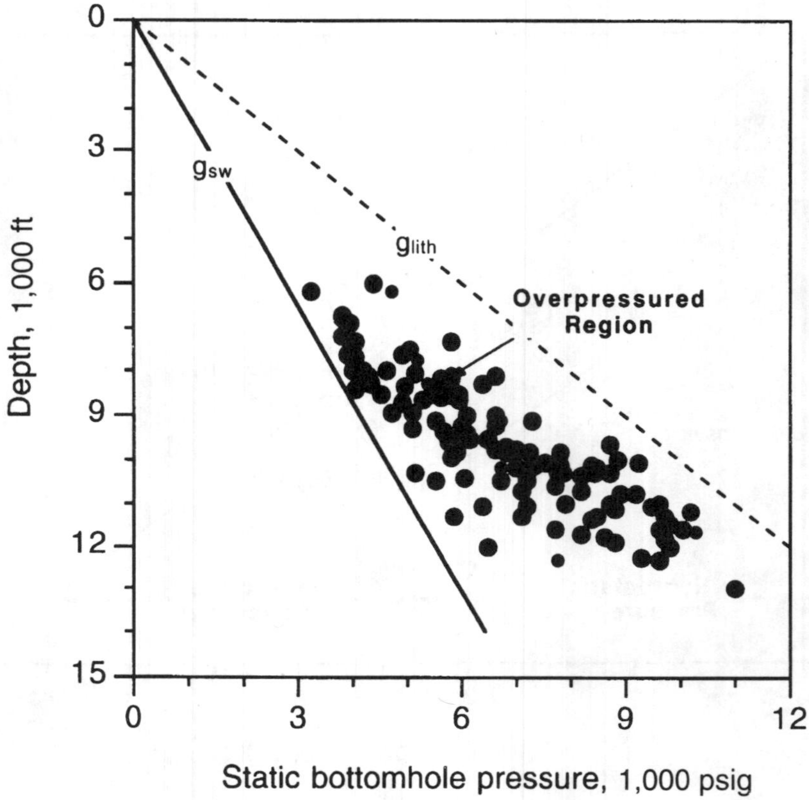


**Fig. 1.2—Measured overpressured zones along the U.S. gulf coast (from Ref. 1).**

Hubbert and Rubey[2] state that the single-valued, total normal stress or lithostatic pressure is the sum of the normal hydrostatic pressure and the effective stress. The term is sometimes defined as the overburden pressure. The summation of stresses may be expressed as

$$\sigma_t = p + \sigma_e , \quad (1.1)$$

where $p$ = hydrostatic or fluid pressure and $\sigma_e$ = effective vertical stress exerted on the rock matrix. Effective stress, often called the net overburden pressure, is the force exerted by the sediment grains through grain-to-grain contacts to help support the lithostatic pressure. Therefore, the total normal stress is the sum of the fluid pressure and the effective stress at the point of measurement. The effective stress, like the total normal stress, cannot be measured directly but is the difference between the total normal stress and the fluid pressure.

*Example Problem 1.1.* A well has been drilled and completed in a gas sand at 10,000 ft. The fluid environment in the sediments is generally salt water, $g_{sw} = 0.465$ psi/ft, while the lithostatic gradient is 1.0 psi/ft. What is the expected net overburden stress at initial conditions?

*Solution.* Rewriting Eq. 1.1 to the following form and solving gives

$$\sigma_e = \sigma_t - p = 10,000(1 - 0.465) = 5,350 \text{ psi}.$$

The net overburden pressure or the stress exerted on the rock matrix is 5,350 psi at initial conditions. This stress will increase as the pore pressure is reduced during production of reservoir gas.

Fluid pressures greater than the normal hydrostatic pressures are called overpressures. The term abnormal pressure is also used to signify overpressured conditions, especially in the U.S. gulf coast basin. However, the adjective abnormal could denote pressures either greater than or less than the normal. Overpressures commonly occur in the deeper subsurface and are preceded by a transition zone of rapid pressure buildup with depth. **Fig. 1.3a** shows that pressures below the transition zone for the overpressured case may follow either a hydrostatic gradient with pressures greater than normal (Line b) or may contine to increase along a gradient that is greater than the normal hydrostatic (Line a).

Overpressures reduce the effective stress, while the total normal-stress gradient remains essentially constant. The result of reduced effective stress is to weaken the overpressured rock. The effective stress approaches zero as fluid pressures increase to near lithostatic pressure. The rock can fail internally by fracture at this point. In other words, the lithostatic pressure is the maximum pressure that the rock can sustain without failure.

Subnormal pressures are present in confined reservoirs in intermontane basins. For example, subnormal pressures may follow the gas gradient. Other areas have subnormal pressures that follow gradients that are less than the normal hydrostatic gradient. These subnormal pressures commonly denote hydrodynamic flow in aquifers that control the pressures in local reservoirs. **Fig. 1.3b** illustrates the underpressured case.Subnormal pressures may increase with depth either parallel to the hydrostatic gradient (Line d) or at lesser gradients (Line c). Drilling problems arise when drilling through these zones because of the potential of lost circulation of the drilling fluids in the low-pressured section.

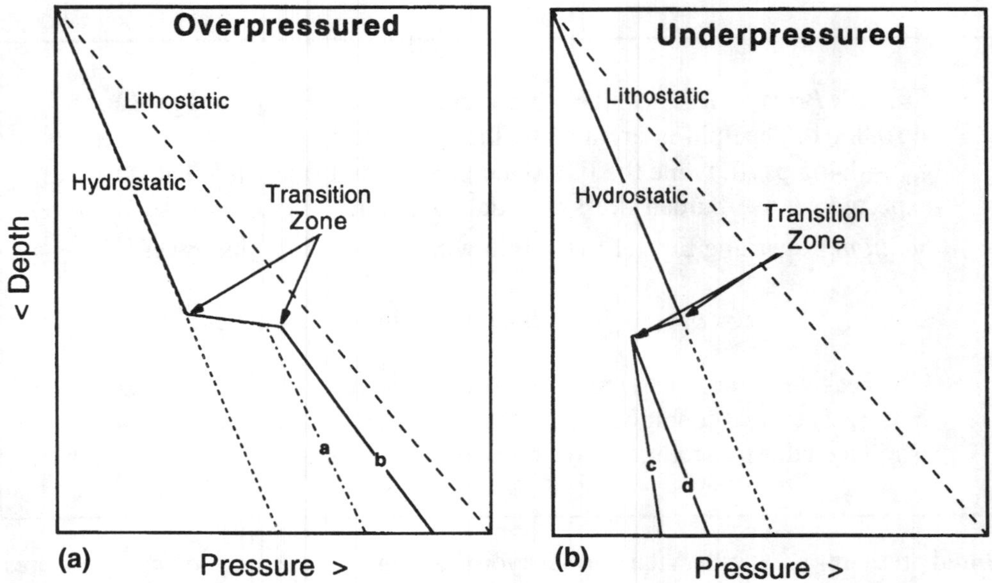


**Fig. 1.3—Diagram of common (a) overpressured gradients encountered in the subsurface and (b) underpressured gradients encountered in the subsurface.**

## 1.2 Detection of Overpressures

Overpressures may be detected by pressure measurements while drilling, during logging, or in the reservoir after the well is cased and completed. Predicting the approximate levels of overpressured zones that the drill bit is expected to encounter is a necessary part of any well program.

**1.2.1 Direct Evidence.** Drillstem tests and repeat formation testers (RFT's) are often used in open and cased boreholes to measure fluid pressures at any particular depth. A drillstem test is conducted by entering the borehole with a closed test tool mounted on the bottom of the drillpipe. A valve located at the bottom of the tool is opened to recover a formation-fluid sample and record the pressure response after the tool is shut in. The tool is equipped with rubber packers that are expanded to seal off the bottom of the hole from the column of drilling mud surrounding the drillpipe before the test is started.

The pressure at the start of the test initially increases rapidly and then increases at a lesser rate until a constant rate of pressure buildup is achieved during the buildup period. The packers are released, and the tool is freed and withdrawn from the borehole after the test procedure is completed. The volumes of oil, gas, and water in the pipe are measured and sampled as the drillpipe is uncoupled at the surface. The flow time of the test is known, so recovered oil, gas, or water volumes can be converted to an average flow rate from the reservoir.

Pressures are recorded during the buildup period by a gauge located either downhole or at the surface. The rate of buildup can be extrapolated to an infinite shut-in time if the constant rate of pressure buildup has been attained.[3,4] This extrapolated pressure represents the original pressure of the reservoir before the withdrawal of fluid during the flow period.

The RFT is similar to the drillstem-test tool. However, it is much smaller and is run into the hole on a wireline. The advantage of the RFT is that several tests may be conducted at different depths in the borehole during a single trip and may be conducted in a much shorter time. The major disadvantages are that the small size of the tool restricts the volume of fluid recovered and the possibility that mud cake might affect the seal of the snorkel to the side of the borehole.

A recording pressure gauge may be lowered on a wireline to the desired depth after a well has been cased, perforated, and completed. The well is shut in, and the buildup pressure is measured as a function of time.

**1.2.2 Indirect Evidence.** Overpressured zones can pose problems during drilling operations if the need for increased mud weight is not anticipated. Predicting the probable depth where high pressures will be encountered in a well scheduled to be drilled is a necessary part of well design. Drilling mud is usually a mixture of fresh water weighing 8.3 lbm/gal and clay particles and chemical additives that raise the mud weight to about 10 lbm/gal and increase the gradient to 0.52 psi/ft . A 10-lbm/gal mud weight is usually sufficient to control normal hydrostatic pressures and prevent flowback into the borehole from aquifers and to minimize borehole caving or enlargement. These conditions are usually called "drilling in a balanced mode."

An underbalanced condition can occur when the bottomhole pressures encountered by the drill bit are greater than the borehole hydrostatic pressure. Drilling problems will not arise if an overpressured section is composed of low-permeability shale because reservoir-fluid feed-in is inhibited by low permeability. However, formation fluids will tend to flow into the borehole if a permeable aquifer is encountered and the in-situ permeability is sufficient to permit significant fluid flow. The flow of formation fluid into the wellbore may be great enough to unload drilling mud from the borehole at the surface. The well will blow out if the blowout preventers fail or have too low a pressure rating to seal off the system from the atmosphere.

Operators have learned to use indirect measurements to detect the approach of overpressured sections. If shale is circulated from the bottom of the hole, densities are often found to decrease as the wellbore approaches an overpressured zone. Shale density usually increases with depth under normal geological conditions. Decreasing shale density can also be detected in the response of well logs. Resistivities of shales normally increase with depth. A decrease in formation density can be detected by a decrease in resistivity or by the conductivity inverse, which will increase. Sonic logs show an increase in interval transit times in overpressured shales. The increased transit times correspond to a density decrease. In fact, log response through overpressured sections can be correlated directly with formation pressure.[5,6] **Fig. 1.4** shows both transient time and shale conductivity increase when an overpressured zone is encountered at 10,000 ft in the Oligocene, Frio in south Texas. In this case the transition from normal to overpressured conditions is quite abrupt. The gradient in the overpressured section varied from 0.85 to 0.88 psi/ft.

Temperatures of overpressured shales sometimes increase. Increasing temperatures may be detected by monitoring change in the temperature of the drilling mud. None of these techniques for interpreting indirect changes in the character of the sedimentary section is completely reliable, but monitoring for these changes during drilling operations can increase the odds of avoiding an expensive accident. Casing can be set when the approach to high pressures is detected. Drilling through the high-pressure zone can then proceed with heavyweight mud without fear of lost circulation in the shallower section.

Formation waters of overpressured sections often show an abrupt decrease in salinity when overpressured zones are encountered; the water changes to slightly brackish from the normally highly saline content. Decreasing shale densities and water salinity suggest that overpressured sections commonly contain shales of low density and water of low salinity relative to their depth of burial. These characteristics, as well as others, indicate the possible origins of overpressures.

## 1.3 Origins of Overpressures

The origins of overpressures are of particular interest when predicting their occurrence, particularly in sparsely drilled basins or in newly developing areas. An understanding of their origin is useful for anticipating potentially harmful effects during drilling and production of overpressured gas or oil reservoirs.

Overpressures have been recorded at depths from a few hundred feet to greater than 25,000 ft and have been reported in numerous areas throughout the world. There are, however, two major settings in which geopressures are principally encountered: (1) recently active tectonic areas of compressional folding and (2) relatively young sedimentary sections in which a great thickness of sediments has been rapidly deposited. The causes of overpressuring, therefore, depend to a great extent on the geologic history of the sedimentary section.

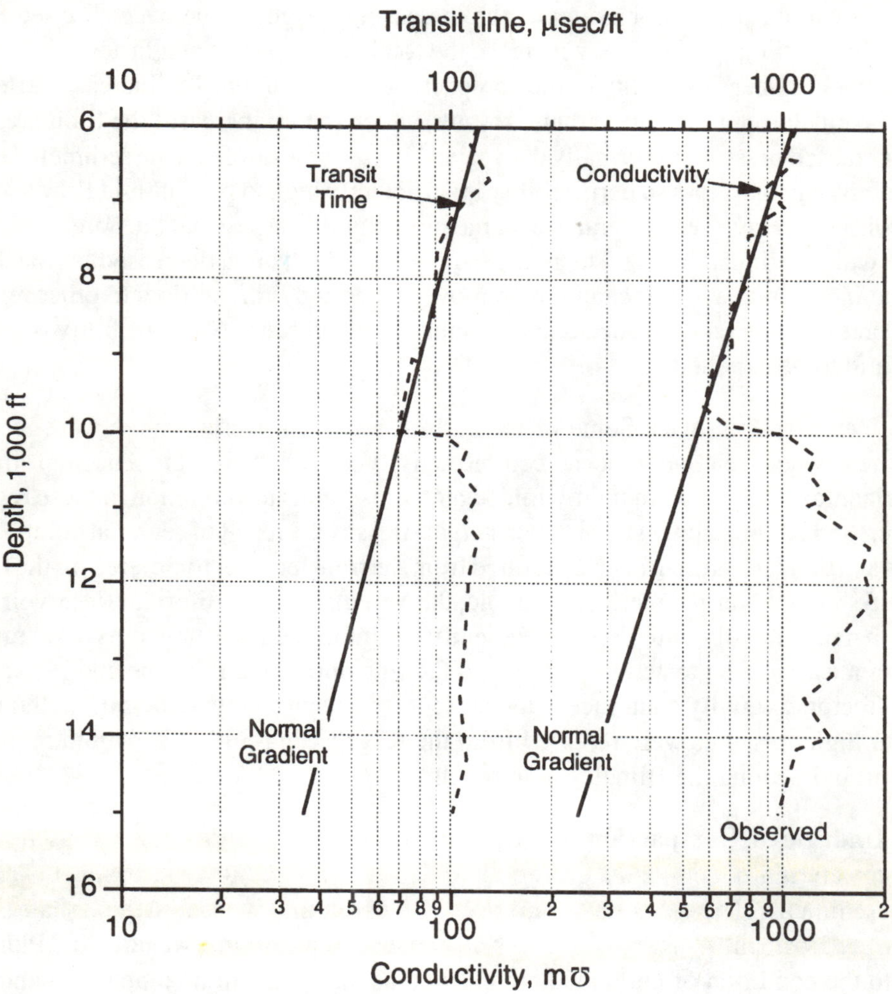


**Fig. 1.4—Shale conductivities and interval transit times through the Frio (Oligocene) in south Texas (adapted from Ref. 7).**

A major requirement for overpressuring is that the sedimentary section consist of rocks of sufficiently low average permeability so that flow is negligibly small compared with the rate of pressure buildup in the sediment. For this reason, overpressures commonly occur in sections composed predominantly of shale.

High fluid pressures result when a portion of the effective stress normally assumed by grain-matrix support is transferred to the intergranular water. This conclusion is derived from the assumed relationship of the fluid pressure, effective stress, and total normal stress shown in Eq. 1.1. For example, the increasing weight of the overlying sediments caused by continuing burial may increase the total normal stress at a faster rate than the increasing load-bearing ability of the compacting porous media. The increased effective stress is transferred to the intergranular water, which in turn increases the fluid pressure. This increased pressure is normally dissipated by flow of water from the sediment at shallow depths when the in-situ permeability is sufficiently high to permit fluid flow to outlying sediments. However, the grains become more tightly packed and the pore space is reduced with continued burial. The reduced pore size results in a decreased permeability, and the intergranular water cannot be removed rapidly by forced flow to other regions. Fluid pressures increase to successively higher levels because of the inability of the excess fluid to be expelled.

**1.3.1 Tectonic Pressures.** Some of the earliest recorded overpressures were reported from areas where relatively recent tectonic activity caused the principal normal stress to be horizontal. The rocks are tightly folded and thrust faults are common in these regions. Hubbert and Rubey[2] suggest that the principal stress is the result of regional compressive forces and that overpressuring has resulted from the transfer of tectonic stress to the fluids. **Fig. 1.5** shows a North American example, the Ventura field, California. Reservoir pressures increase rapidly with depth until nearly attaining the lithostatic pressure gradient. This area was subject to active folding and faulting during the Pliocene and Pleistocene Ages. Tectonic activity continues to the present, as evidenced by numerous earthquakes. Similar high pressures were reported from the Khaur, Chia-Surkh, and Qum regions in Iraq and India along the Himalayan fold belt.

**1.3.2 Undrained Compaction.** Dickinson[1] recognized overpressures as the result of slow dewatering of sediments during rapid deposition and burial in basins containing young sediments. The common occurrence of overpressures in the first thick shale underlying a predominantly sandy section is the evidence for this origin. Magara[8] and Plumley[9] refer to the condition of undrained compaction as disequilibrium compaction and nonequilibrium compaction, respectively. However, the term undrained compaction seems to describe the lack of normal compaction and dewatering of the section.

Tertiary sediments of the Gulf of Mexico basin provide typical examples of undrained compaction. **Fig. 1.6a** presents the sand/shale relationship through the sedimentary section from 2,500 to 16,000 ft in the McAllen Ranch field, south Texas. **Fig. 1.6b** represents fluid pressures derived from conductivity measurements. The normal saltwater gradient of 0.465 psi/ft and a 0.9-psi/ft gradient are included to highlight the location of the overpressured zones. The first overpressured zone is in the Frio formation near 5,000 ft, while the second is in the deeper Vicksburg formation beginning at 7,500 ft. Both sections composed predominantly of shale are interpreted to be abnormally pres-

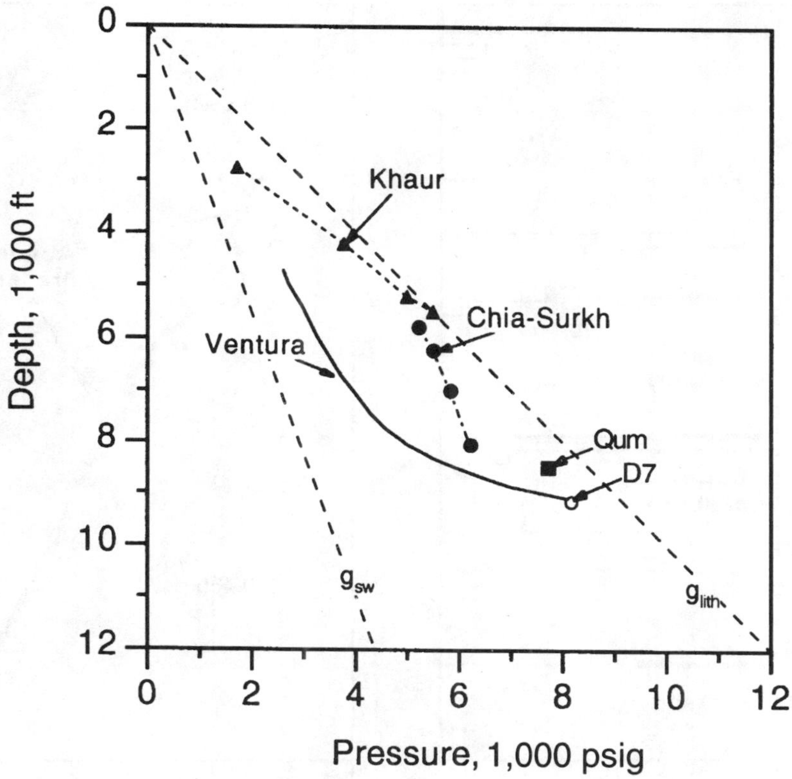


**Fig. 1.5—Overpressures recorded in wells drilled in or near active tectonic belts of compressional folding and faulting (from Ref. 2).**

sured. An intervening section of near-normal pressures occurs in the sandstones of the lowermost Frio. There is good agreement between the calculated pressures and the single measured pressure at 11,850 ft.

The upper, overpressured zone probably results from undrained compaction in the shallow Frio formation. **Fig. 1.7** shows that the densities calculated from a density log are considerably less in the overpressured sections than those observed in the normal-pressured section. A density expected for normal compaction conditions is included for comparison. The Frio shales display exceedingly low densities when compared with this curve. Shale densities are much greater in the normally pressured sandstones of the basal Frio encountered at approximately 14,000 ft. A low shale density is also shown in the deeper overpressured zone. Core- and log-derived density values are included in the figure for comparison purposes.

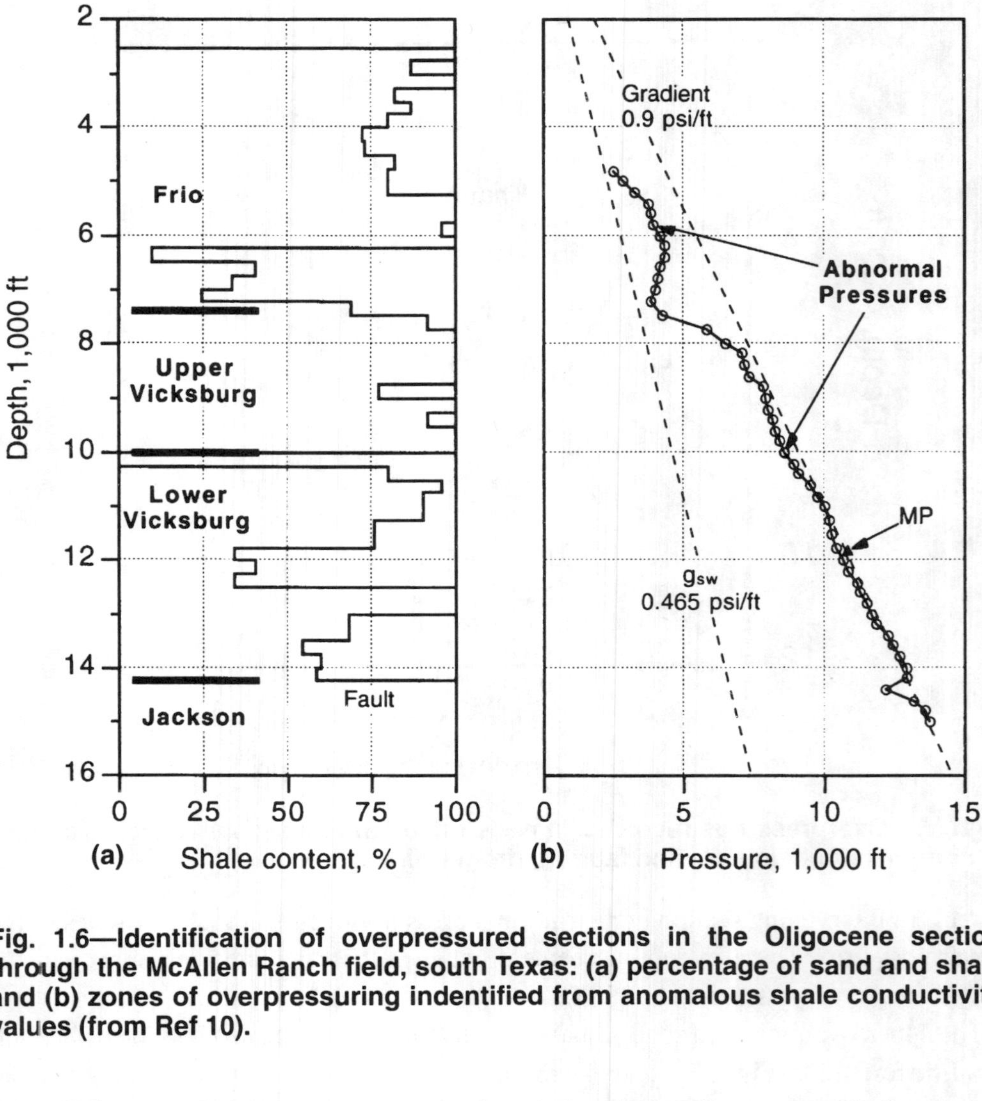


**Fig. 1.6—Identification of overpressured sections in the Oligocene section through the McAllen Ranch field, south Texas: (a) percentage of sand and shale and (b) zones of overpressuring indentified from anomalous shale conductivity values (from Ref 10).**

**1.3.3 Clay Transformation.** In many basins, shales are usually composed of a mixture of smectite, illite, and kaolinite clay minerals. Smectite clays contain as many as four layers of water molecules within their crystalline structure. They are collectively referred to as expandable clays because of their ability to absorb water. Absorbing water within the crystalline structure causes the clay particle to swell, increasing the apparent grain volume. An example of an expandable clay is montmorillonite formed by weathering of vol-

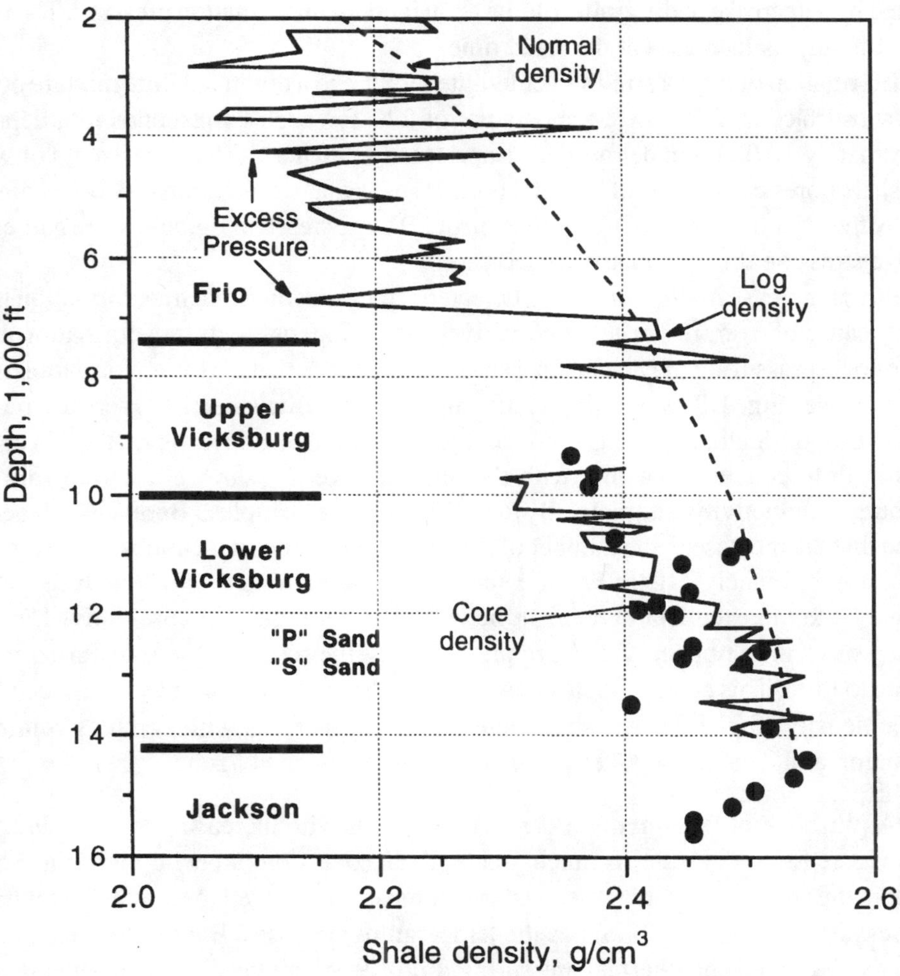


**Fig. 1.7—Shale densities in the Oligocene Frio and Vicksburg formations, McAllen Ranch field, south Texas (from Ref. 10).**

canic ash to bentonite. More properly, these expandable clays are now known as interlayered smectite/illite.

Smectite/illite clays found at shallow depths and at low temperatures typically contain largely smectite with its interlayered water. Increased temperature causes one or more of the water layers to be expelled. The interlayer spacing is reduced to the structure of illite.[11,12] The expelled water occupies an increased pore volume (PV) that must be accommodated by a decrease in the grain volume or an increase in formation pressure. The number of illite layers increases at the same time.

Transformation of interlayered smectite/illite mixture to a dominant illite mixture occurs in the subsurface at an average temperature of 220°F, which corresponds to a depth of approximately 10,000 ft under normal temperature gradients.[11] Thus, the excess of water in the shale pores carries part of the total normal stress that would ordinarily be accommodated by the effective stress of the porous media. This increase in fluid pressure causes the effective stress on the rock matrix to decrease.

An abrupt increase in illite should be detected if the smectite/illite transformation is the primary cause of overpressuring. Alternatively, the temperature of transformation at the depth of overpressuring may be taken as indirect evidence that clay transformation is the primary cause. **Fig. 1.8,** where the relative numbers of illite/smectite layers are plotted as a function of depth, applies this concept to the McAllen Ranch field. The abrupt increase in illite content occurring with the onset of overpressures is inferred from the Vicksburg conductivity anomaly. By analysis of core samples, Berg and Habeck[10] showed that the increase in the number of illite layers coincides with a subsurface temperature of approximately 212°F; Freed[13] showed the same thing through analysis of drill cuttings. Note the excellent agreement between the cuttings- and core-derived values. Overpressures in the upper Vicksburg are probably the result of the transformation of smectite to illite. This change in clay composition also coincides with an abrupt decrease in shale density (Fig. 1.7). The shale densities again increase with depth as continued compaction and dewatering takes place below this low-density zone.

**1.3.4. Aquathermal Pressuring**. Pore water expands with increasing temperature. The volume increase of the water is much greater than expansion of clay minerals normally present in the shales. If the increased water volume is not allowed to escape, the total normal stress increases during burial as the temperature increases. Barker[14] states that this mechanism is called aquathermal pressuring. **Fig. 1.9,** where the change in water density expressed in g/cm$^3$ units is plotted as a function of temperature and pressure, illustrates the physical process. The various density lines represent changes for a water characterized by a particular initial density. The increasing temperature scale shown on the vertical axis also represents increasing depth of burial of the zone of interest because formation temperature is a function of depth. The density of the water decreases because of thermal expansion. The increased water volume caused by thermal expansion may be accommodated in the reservoir by two different means: (1) the expanding water is able to escape from the shale if the permeability is high enough to permit fluid flow to occur or (2) pressure buildup will occur when the permeability in the shale is very low and all the expanded water volume is retained.

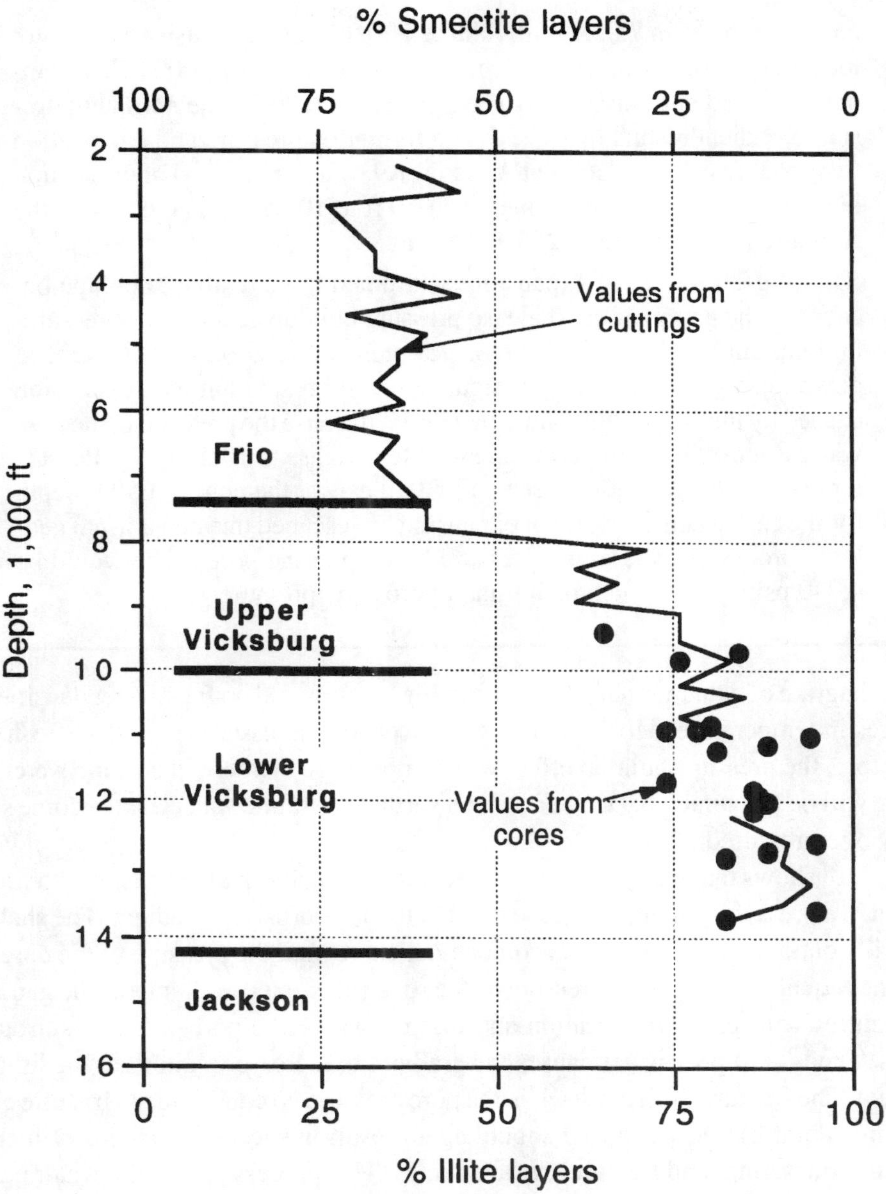


**Fig. 1.8—Relative numbers of smectite and illite layers in the interlayered clay minerals, McAllen Ranch field, south Texas. Analyses of drill cuttings (from Ref. 13) and of core samples (from Ref. 10).**

*Example Problem 1.2.* A unit volume of 0.975-g/cm$^3$ density pore water located in an impermeable shale is buried to a depth of 10,000 ft. The temperature gradient is assumed to be $g_T = 1.37°F/100$ ft. The normal hydrostatic gradient results in a 4,330-psig formation pressure and a formation temperature of 202°F at Point A in Fig. 1.9. Continual burial of the section increases the depth of the zone by 3,650 ft to 10,650 ft and increases the formation temperature to 250°F. Explain.

*Solution.* The water of expansion is assumed to be unable to escape because of the nature of the shale, so pressure buildup occurs. The pressure-buildup curve will parallel the 0.975-g/cm$^3$ correlation line and extend down to 250°F to Point B. The figure indicates that volumetric expansion caused by the 50°F temperature increase will cause the pressure in the pore water trapped in the impermeable rock to increase to 10,650 psig. Point C represents a hydrostatic pressure of 5,910 psig at the new 13,650 ft depth for the case when the water of expansion has escaped through normal geologic processes. Therefore, pressure buildup in the pore space would be 4,740 psig greater than the normal hydrostatic pressure.

The diagram of aquathermal pressuring in Fig. 1.9 is based on freshwater densities with changes in temperature. More-saline waters expand to a lesser degree than fresh water. Therefore, the pressure-buildup effect with temperature increase in a saline water atmosphere will be at a somewhat slower rate. Nevertheless, saline waters can become significantly overpressured.

**Fig. 1.10** shows that the changes observed during aquathermal expansion should be reflected in a pressure-gradient increase parallel to the hydrostatic gradient. The shale density will continue along the normal compaction curve until the pressure of the pore water becomes equal to the fracture pressure of the rock (illustrated at Point a). The generation of fractures will result in migration and loss of part of the pore water to surrounding shales. The loss of pore water causes the gradient to increase to intersect the lithostatic gradient. The fractures will close when the pore pressure is reduced to the fracture closing pressure (Point b). Deeper burial should again result in successive cycles of increased pressure, fracturing, and fracture-closure cycles. This process causes the water pressure to be maintained essentially equal to the fracture pressure.

Barker and Horsfield[15] have proposed that aquathermal pressuring has been active mechanism in the Tertiary section of the U.S. gulf coast province but have not been completely successful documenting these effects. Berg and Habeck[10] believe aquathermal

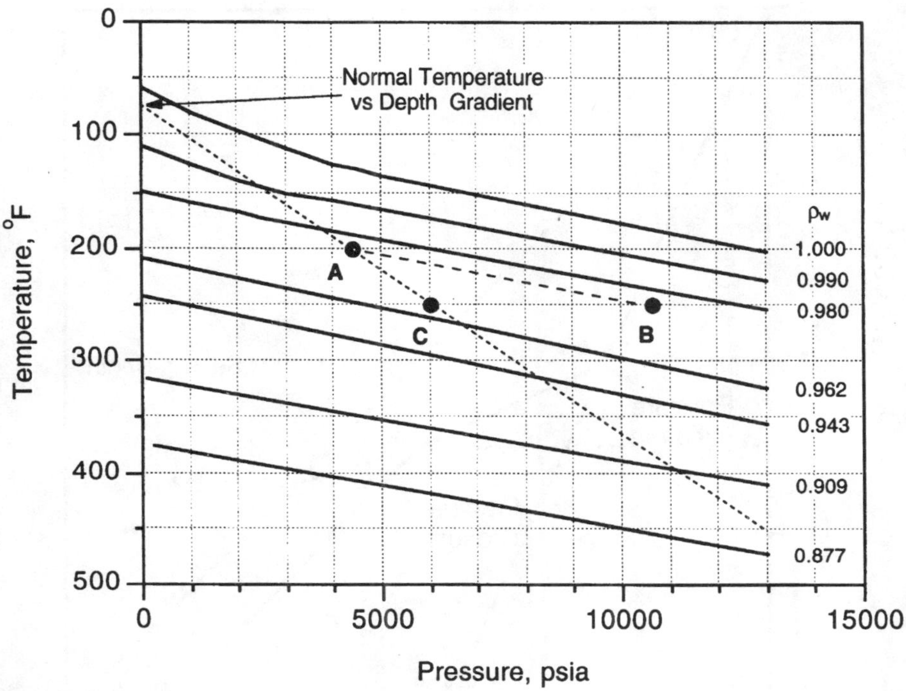


**Fig. 1.9—Diagram of pressure exerted by the expansion of a unit volume of water during burial and a 50°F temperature increase (from Ref. 14).**

pressuring may be taking place, for example, in Oligocene Vicksburg shales of south Texas. However, shale densities do not remain constant but are reduced through the interval of pressure buildup, thereby casting doubt on the validity of the mechanism.

**1.3.5 Oil and Gas Generation.** The conversion of solid organic matter, or kerogen, to liquid or gaseous petroleum also results in a volume increase. The newly generated oil or gas exerts a pressure by fluid expansion much like that proposed for aquathermal-pressuring process. An increase in pressure can also result in fracturing of the source rock. The induced fractures assist in the primary migration of oil or gas from source rock to carrier beds. Gas generation is more likely to cause overpressuring because of its greater ability to expand compared with oil. The process has been applied to the thermal degradation of oil to gas as the result of deep burial and high temperatures.[16] The same principles apply equally well to the generation of gas from kerogen.

**Fig. 1.11** illustrates the process. Fig. 1.11a shows that a unit PV contains both oil and water in the stage of initial saturation. Deeper burial with a resultant higher temperature

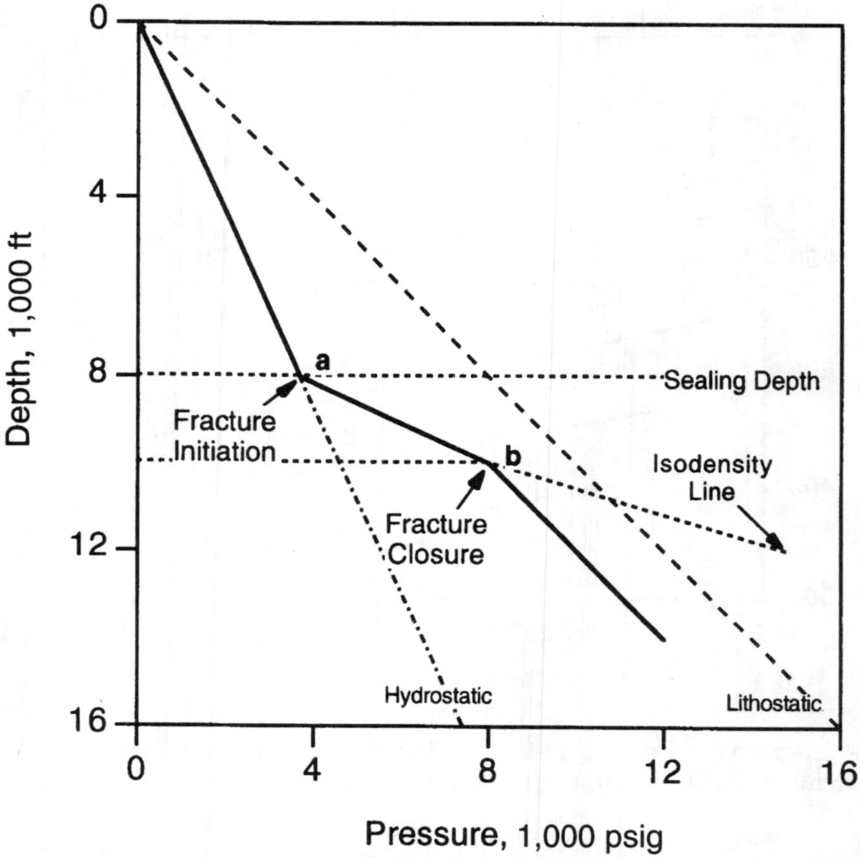


**Fig. 1.10—Diagram of pressure buildup by aquathermal pressuring caused by burial and increased temperature.**

converts a portion of the oil to gas (Fig. 1.11b). A small amount of carbon residue remains. The oil, gas, and water expand at this higher temperature, and some of the gas goes into solution in the water and oil. The result is an increase in pressure because of fluid expansion. Eventually all the oil is converted to gas at a still deeper level (Fig 1.11c). An increased amount of residue is left at this last stage, but the pressure exerted by the gas and water is at a higher level. The increased fluid volumes exert pressures that reach the fracture pressure of the rock after conversion of only 25 to 50% of the initial oil saturation if the rock acts as an impermeable barrier.[16] Kerogen can be substituted for the initial and reduced oil saturations, and the effect of gas and water expansion would be essentially the same for this sequence.

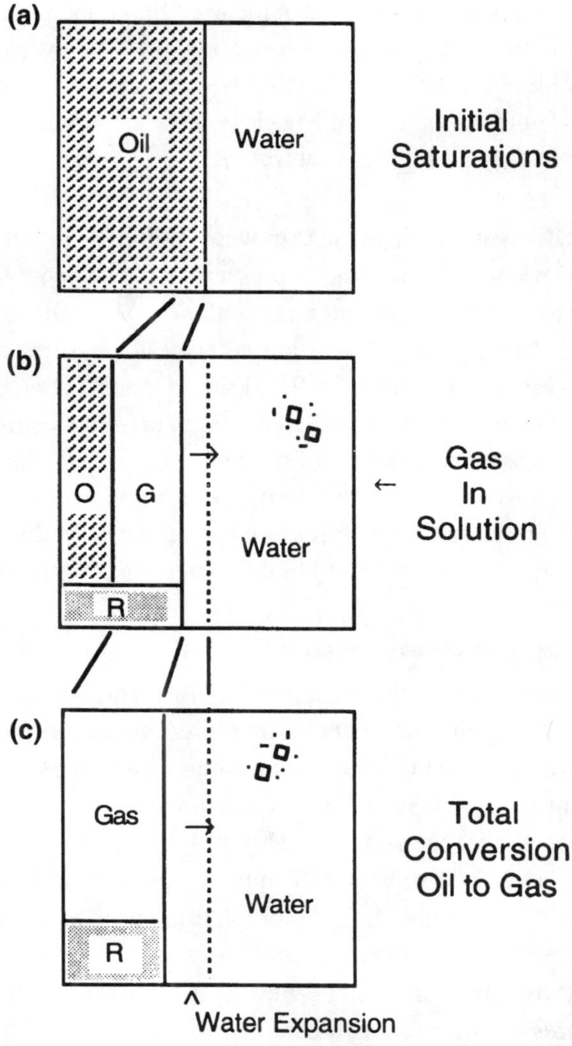


**Fig. 1.11—Conversion of oil to gas in a unit PV during burial and increase in temperature because of fluid expansion (from Ref. 16).**

Pressure buildup during the gas-generation process could be rapid and eventually reach the fracture pressure of the rock in much the same manner as visualized for aquathermal pressuring. There is a slight reduction in pressure after fracturing occurs. These fractures close with increased depth and temperature. Shale densities are only slightly reduced by fractures and should remain essentially constant over the interval of pressure buildup.

The deepest overpressures observed at the McAllen Ranch field are probably the result of gas generation. Pressure gradients exceed 0.90 psi/ft below a depth of 12,000 ft. These pressures coincide with the generation of wet gas in the underlying Jackson shales.[17] These shales (seen at the bottom of Fig. 1.7) show a decreasing density with depth that is probably the result of a high gas saturation.

**1.3.6 Osmotic Pressures.** Differences in water salinities are known to occur in the subsurface. Jones[18] proposed that osmotic pressures may be generated by flow of ions from the more saline to the less saline waters. Calculations show that osmotic-pressure buildup could result in overpressuring. For example, excess pressures could be on the order of 1,500 psi with a salinity difference of 100,000 ppm NaCl between two formations. These pressures, however, would not be sufficiently high to produce gradients in excess of about 0.6 psi/ft that are common at depths of overpressuring. A further objection is that higher osmotic pressures would be on the side of more saline waters whereas reduced salinity is commonly observed in the overpressured zone. The conclusion is that osmotic pressures would not greatly contribute to higher levels of overpressures.

## 1.4 Preservation of Overpressures

Two views exist concerning the preservation of overpressured conditions in sediments. These opposing viewpoints are that overpressures are preserved as a static condition and that overpressures are a dynamic condition when viewed in geologic time and are a product of several influences. How have overpressures observed in older sediments, such as those of Cretaceous Age, been preserved for millions of years? The compartments of deep sediments must be sealed by absolutely impermeable barriers to maintain high pressures for long periods of time for the static condition to occur.[19] Such a view holds that the overpressure regime is developed from a single origin.

Alternatively, overpressures can be viewed as a dynamic condition resulting from one or more pore-pressure-increasing mechanisms operating at different levels and for different reasons. For example, undrained compaction occurs in all sediments at shallow depths during the dewatering process. Low-permeability relatively impermeable shales become overpressured, while the adjacent sandstones remain normally pressured because of their greater propensity to transmit water. The section is subjected to increased temperature with burial and subsidence to greater depths. Dehydration of expandable clays allows a second dewatering stage to take place. Finally, at great depths, the organic matter in the shales reaches maturity and oil and gas are generated as a function of temperature and time.

## Nomenclature

$D$ = depth, L, ft
$g$ = gradient, m/t$^2$, psi/ft
$g_{fw}$ = freshwater gradient, m/t$^2$, psi/ft
$g_{lith}$ = lithologic gradient, m/t$^2$, psi/ft
$g_o$ = oil gradient, m/t$^2$, psi/ft
$g_{sw}$ = saltwater gradient, m/t$^2$, psi/ft
$g_T$ = temperature gradient, T/L, °F/ft
$p$ = pressure, m/Lt$^2$, psi
$p_h$ = hydrostatic pressure, m/Lt$^2$, psi
$\rho_b$ = bulk density of reservoir-rock material, m/L$^3$, lbm/ft$^3$
$\rho_w$ = water density, m/L$^3$, lbm/ft$^3$
$\sigma_e$ = effective stress, m/Lt$^2$
$\sigma_t$ = total normal stress, m/Lt$^2$

## References

1. Dickinson, G.: "Geological Aspects of Abnormal Reservoir Pressures in Gulf Coast Lousiana," *AAPG Bull.* (Feb. 1953) **37,** No. 2, 410.
2. Hubbert, M.K. and Rubey, W.W.: "Role of Fluid Pressure in Mechanics of Overthrust Faulting, I. Mechanics of Fluid-Filled Porous Solids and Its Application to Overthrust Faulting," *Bulletin,* Geological Soc. of America (Jan. 1959) **70,** No. 1, 115.
3. Horner, D.R.: "Pressure Buildup in Wells," *Proc.,* Third World Petroleum Congress, E.J. Brill (ed.), Leiden, Holland (1951) Sec. II, 503.
4. Lee, W.J.: *Well Testing,* Textbook Series, SPE, Richardson, TX (1982) **1,** 159.
5. Hottman, C.E. and Johnson, R.K.: "Estimation of Formation Pressures From Log-Derived Shale Properties," *JPT* (June 1965) 717; *Trans.*, AIME, **234.**
6. Palicaukas, V.V. and Domenico, P.A.: "Microfracture Development in Compacting Sediments: Relation to Hydrocarbon-Maturation Kinetics," *AAPG Bull.* (June 1980) **64,** No. 6, 927.
7. Matthews, W.R. and Kelley, J.: "How to Predict Formation Pressure and Fracture Gradient from Electric and Sonic Logs," *Oil & Gas J.* (20 Feb. 1967) 92.
8. Magara, K.: "Importance of Aquathermal Pressuring Effect in Gulf Coast," *AAPG Bull.* (Oct. 1975) **59,** No. 10, 2037.
9. Plumley, W.J.: "Abnormally High Fluid Pressure," *AAPG Bull.* (March 1980) **44,** No. 3, 414.
10. Berg, R.R. and Habeck, M.F.: "Abnormal Pressures in the Lower Vicksburg, McAllen Ranch Field, South Texas," *Trans.*, Gulf Coast Assn. of Geological Societies (1982) **32,** 247.
11. Burst, J.F.: "Diagenesis of Gulf Coast Clayey Sediments and Its Possible Relation to Petroleum Migration," *AAPG Bull.* (Feb. 1969) **53,** No. 1, 73.
12. Powers, M.C.: "Fluid Release in Compacting Marine Mudrocks and Their Importance in Oil Exploration," *AAPG Bull.* (July 1967) **51,** No. 7, 1240.

13. Freed, R.L.: "Shale Mineralogy and Burial Diagenesis of Frio and Vicksburg Formations in Two Geopressured Wells, McAllen Ranch Area, Hildalgo County, Texas," *Trans.*, Gulf Coast Assn. of Geological Societies (1981) **31,** 289.
14. Barker, C.: "Aquathermal Pressuring—Role of Temperature in Development of Abnormal-Pressure Zones," *AAPG Bull.* (Oct. 1972) **56,** No. 10, 2068.
15. Barker, C. and Horsfield, B.: "Mechanical Versus Thermal Cause of Abnormally High Pore Pressures in Shales," *AAPG Bull.* (Jan. 1982) **66,** No. 1, 99.
16. Barker, C.: "Calculated Volume and Pressure Changes During the Thermal Cracking of Oil to Gas in Reservoirs," *AAPG Bull.* (Oct. 1990) **74,** No. 10, 1254.
17. Tanner, J.A.: "Chemical and Isotropic Evidence of the Origin of Hydrocarbons and Source Potential of Rocks from the Vicksburg and Jackson Formations of Slick Ranch Area, Starr County, Texas," MS thesis, U. of Houston, Huston (1987) 289.
18. Jones, P.H.: "Hydrology of Neogene Deposits in the Northern Gulf of Mexico Basin," *Proc.*, First Symposium on Abnormal Subsurface Pressure, R.E. Ferrell Jr. and B.R. Hise (eds.) (1967) 91.
19. Hunt, J.M.: "Generation and Migration of Petroleum from Abnormally Pressured Fluid Compartments," *AAPG Bull.* (Jan. 1990) **70,** No. 1, 1.

## SI Metric Conversion Factors

| | | | |
|---|---|---|---|
| ft | $\times$ 3.048* | E − 01 | = m |
| °F | (°F − 32)/1.8 | | = °C |
| gal | $\times$ 3.785 412 | E − 03 | = $m^3$ |
| lbm | $\times$ 4.535 924 | E − 01 | = kg |
| psi | $\times$ 6.894 757 | E + 00 | = kPa |

*Conversion factor is exact.

Chapter 2

# Effect of Overpressures on Reservoir-Rock Properties

Reservoir sandstones located in overpressured sections commonly show increased porosity when compared with normal-pressured sections of similar rock character. Increased porosity is accompanied by changes in other reservoir properties, such as increased permeability and oil and gas volumes and decreased capillary displacement pressures. The increased porosity is largely the result of the dissolution effect of flow of formation waters over geologic time. Overpressuring of any section means that the potential for flow and the release of overpressure can take place by flow through sandstones and/or along fractures or faults. Primary porosity is a product of the deposition environment as well as of the normal compaction process and generally decreases with depth as a continuous function. The increased porosity at depth caused by dissolution is known as secondary porosity.

## 2.1 Primary Porosity

The general sequence of sandstone-porosity reduction with depth can be illustrated diagrammatically by two different rock compositions commonly found in hydrocarbon reservoirs: quartzose and lithic sandstones. **Fig. 2.1** illustrates the two types. The general sequence of pore-volume (PV) alteration is compaction, cementation, dissolution, and finally, recementation.

Quartzose sandstones are those with a dominance of quartz grains, generally >70%. The remaining portion of the grain composition is feldspars, rock fragments, and/or clay minerals.[1] Quartz is resistant to mechanical deformation and forms an aggregate with a relatively rigid framework. Porosity reduction with depth takes place first by compaction. Mechanical adjustments between grains result in a tighter packing of the grains. The early period of compaction is followed by increasing amounts of cementation as compaction waters from the enclosing shales flow through the sandstones. Quartz is commonly precipitated between the grains and appears as overgrowths on the quartz grains. Porosity reduction with depth is essentially a linear function for quartzose sandstones.

In contrast, lithic sandstones generally exhibit a low quartz content, usually <40%. The balance of the grains is a combination of feldspars, lithic rock fragments, and clay minerals. The lithic component generally consists of fine-grained rock fragments, such as shales or volcanic rocks. A rigid framework does not exist in lithic sandstones. Compac-

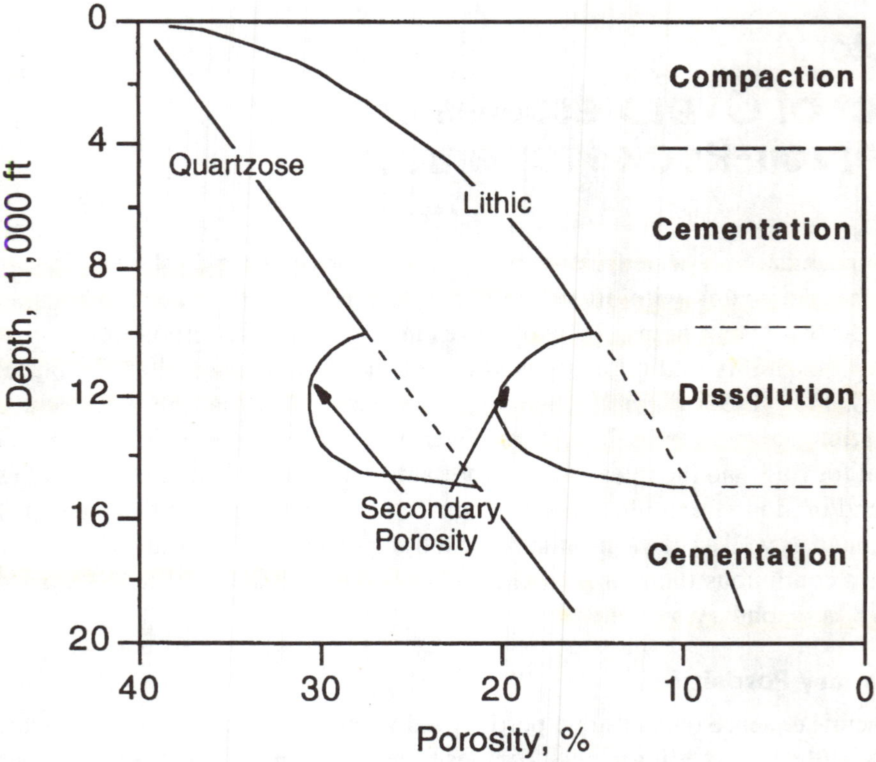


**Fig. 2.1—General sequence of burial diagenesis for quartzose and lithic sandstones.**

tion takes place largely by distorting and squeezing the lithic particles into the interstitial pore spaces. Porosity is reduced at a greater rate than that observed for a quartzose sandstone. This stage of shallow compaction is followed by precipitation of calcite cement, which further reduces porosity. The net effect is that compaction of lithic sandstones follows a curved function, much like that for shale compaction.

The linear porosity reduction of quartzose sandstones with depth for Miocene gas reservoirs in south Louisiana is well documented. Atwater and Miller[2] provide a number of illustrations, and Berg and Habeck[3] discuss an example of lithic sandstones with the characteristic nonlinear porosity reduction with depth. The Oligocene sandstones of south Texas contain a large component of volcanic-rock fragments. The near-surface initial porosity for both types of sandstones is approximately 40%. Porosity is reduced with continued burial in a normal-pressured section until a minimum value of $<10\%$ is reached at great depth. The sandstones are tightly cemented at this point.

## 2.2 Secondary Porosity

Many overpressured sandstone sections contain secondary porosity. The increased PV is greater than would be expected if normal compaction and cementation processes had taken place. The secondary porosity can be the result of dissolution of either grains or cements. Ordinarily, orthoclase potassium feldspars have an open framework along which dissolution has taken place parallel to the cleavage planes. This type of dissolution results in an increased porosity on a microscale. In other cases, feldspar dissolution can result in removal of entire grains, leaving oversized pores with respect to pores located in the surrounding rock.

Dissolution of calcite cement is another common form of secondary porosity. Evidence for dissolution is in the form of cement remnants along the surface of other grains or, in some cases, the occurrence of oversized pores between more rigid grains in an otherwise well-compacted rock. Dissolution of sodium and calcium feldspar plagioclase and volcanic-rock grains have also been observed.

The formation of secondary porosity by dissolution of grains or cement can result in 10 to 15% more porosity added to an otherwise well-consolidated sandstone. Below the dissolution stage, the rock generally again begins to lose porosity by further cementation at increased depths until the minimum porosity is reached. Porosity can be reduced by pressure solution and interpenetration of grains until the pores are reduced to thin slots at the greatest depths.

The level of dissolution and secondary porosity is generally at or near the level of overpressuring and commonly occurs at a depth of about 10,000 ft. However, secondary porosity may occur above the overpressure levels because of upward flow of the groundwater and sometimes at greater depths as overpressuring reaches deeper levels. The depths shown on the burial diagram in Fig. 2.1 depend on the occurrence of abnormally high pressures and flow paths through the section. Fig. 2.1 is for illustrative purposes only because the depth of each porosity alteration does not represent absolute values for any particular area.

**2.2.1 Causes of Dissolution.** The composition of formation waters is generally greatly altered from the sea water that originally saturated the rocks. These connate waters contain acidic ions that are capable of dissolving various mineral grains and cements. Morton and Land[4] identified three types of formation waters that have been recognized in the Cenozoic sediments of the U.S. gulf coast: sodium acetate ($NaC_2H_3O_2$) water, NaCl-rich water, and calcium-rich water .

$NaC_2H_3O_2$ water is the most abundant and contains carboxyl-type ($CH_3COOH$) organic acids, which are produced during generation of oil and gas from organic matter. This water is more dilute than sea water and might have been derived from the loss of interlayer water from clays that were also petroleum source rocks. NaCl water is the next most abundant type and is generally highly saline. This water might have been $NaC_2H_3O_2$ or meteoric water that dissolved halite when in contact with salt diapirs. Calcium-rich water is the least abundant and probably represents water from deeper Mesozoic rocks that has migrated upward into younger sediments. It contains HCl and is relatively acidic. Land and MacPherson[5] state that the excessive calcium might have been derived from dissolution of plagioclase feldspars.

The amount of mineral dissolution observed in deep reservoirs means that large volumes of formation waters have moved through the sandstones in response to pressure differences. In many cases, it is difficult to account for the mass balance between the original composition of the rocks, the apparent volumes dissolved, and the required volumes of transmitted water. Nevertheless, the evidence for dissolution is sound and is based on observation in numerous geologic sections.[6]

**2.2.2 Secondary Porosity in Lithic Sandstones. Fig. 2.2** shows an example of secondary porosity in a portion of the Oligocene, Frio-Vicksburg section in south Texas. The reservoir sandstones contain a relatively low quartz content of about 25% and relatively abundant feldspars, volcanic fragments, and calcite cement. The Vicksburg is overpressured below a depth of about 8,000 ft, with observed borehole gradients $>0.85$ psi/ft[3].[3]

A normal compaction curve for the sandstones is inferred from the decrease in the average interval sonic-log transit times and from a single core with a measured minimum porosity of 9% and no visible secondary porosity present in the thin-section. Other cores throughout the highly pressured section show evidence of secondary porosity by dissolution. Measured porosities increased to as much as 24%.[7] In fact, the difference between the normally expected porosity and the measured porosity is essentially equal to the number of secondary pores observed in thin-sections. Porosity is reduced to normal values of 8% at depths near 15,000 ft, which represents the level of a major listric normal fault. A study of reservoir pressures shows that the listric fault represents a path for water to flow out of the overlying highly pressured section.[3] Presumably, flowing formation water dissolved both grains and cements while moving through the sandstones. The dissolved minerals were then precipitated along the fault. Note, the similarity of Fig. 2.2 with the idealized diagram in Fig. 2.1.

**2.2.3 Secondary Porosity in Quartzose Sandstones. Fig. 2.3** shows a second example of high porosity in a deep section of the Upper Cretaceous Tuscaloosa in south Louisiana. The gas-bearing reservoirs occur in a thick sedimentary sequence that is highly overpressured. Borehole-pressure gradients of 0.6 to 0.9 psi/ft are common. The sandstones are quartzose and either uncemented and extremely porous or well cemented by silica. Fig. 2.3a compares the average Miocene and updip Cretaceous porosity/depth trends with

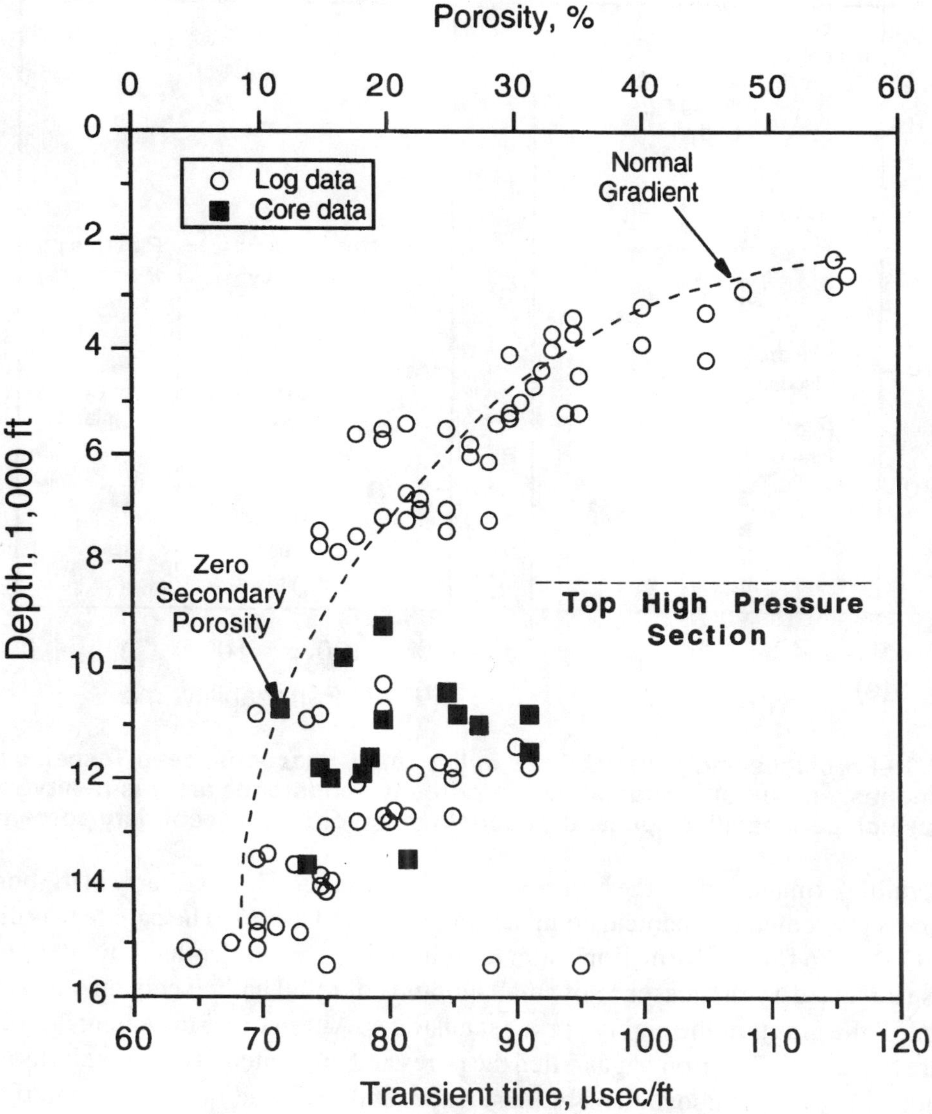


**Fig. 2.2—Porosity as a function of depth in the McAllen Ranch field, south Texas (from Ref. 3).**

those of the deeper Tuscaloosa, Port Hudson, Profit Island, and False River fields. Thin sandstones with silica cement display primary porosities close to 10%, whereas the secondary porosities of those without cement are in the range of 20 to 27%.

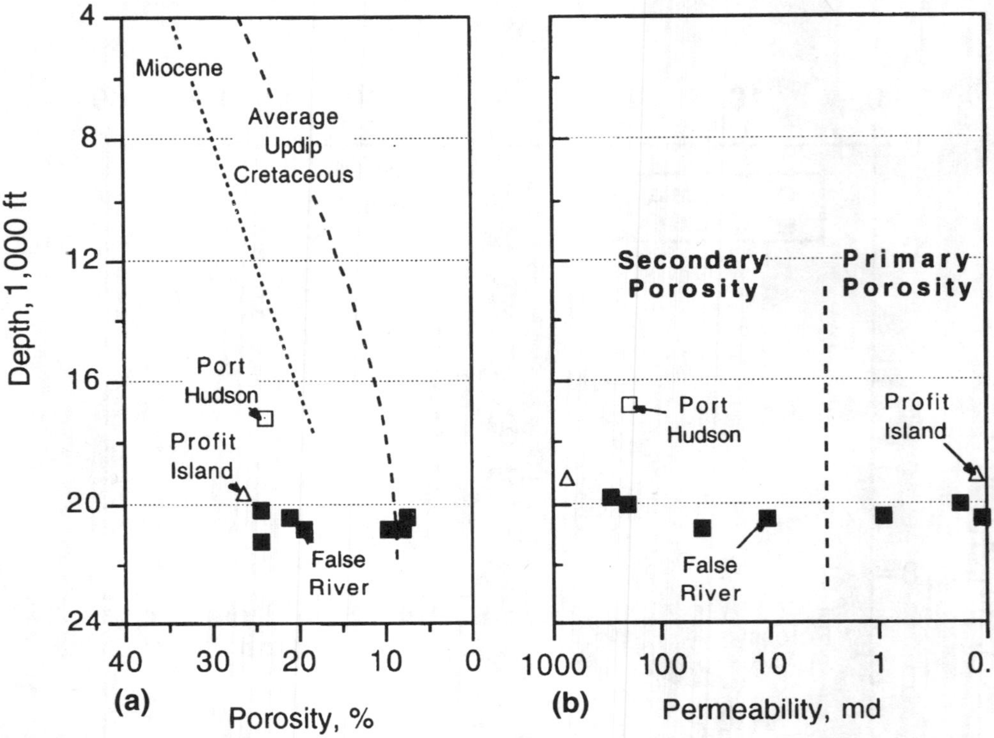


**Fig. 2.3—Porosities and permeabilities in the Late Cretaceous, deep Tuscaloosa sandstones, south Louisiana: (a) higher-porosity sandstones are gas reservoirs and (b) high permeabilities generally occur in reservoirs with secondary porosity.**

A possible explanation for the high porosities is that calcite formed early in the burial process prevented mechanical compaction of the sand grains. The calcite was dissolved by updip flow of formation water after burial to greater depths. This explanation is suggested by the presence of small amounts of relict calcite cement adhering to some of the grains of the highly porous sandstones. Alternatives to cement dissolution are that early migration of gas filled the pores and prevented calcite precipitation or that chlorite clays deposited on grain surfaces and inhibited further cementation. Neither of these ideas is likely in view of the pervasive calcite that now tightly cements the low-porosity sandstones.

The secondary porosity is related to depositional facies. The porous sandstones are coarse grained and were deposited in deep-sea channels. Roth[8] states that the low-porosity sandstones are fine grained and were deposited as an overbank sediment. It appears that formation water flowed updip through the coarse channel sandstones, possibly because of the higher initial porosities. Leaching of the calcite cement during flow greatly enhanced the porosity of the channel sands.

## 2.3 Secondary Permeability

Increased permeability caused by the formation of secondary porosity can be called secondary permeability. Fig. 2.3b illustrates the effect of increased porosity on permeability in the deep Tuscaloosa sandstones. Permeabilities in the tightly cemented, low-porosity sandstones are in the range of 0.1 to 1.0 md, whereas permeabilities of the high-porosity sandstones range from 10 to 700 md. Well productivity is much greater in the sections with secondary permeability.

## 2.4 Layered Reservoirs Caused by Differential Dissolution

Secondary porosity and permeability can be pervasive throughout thick sandstone sections; however, individual bed properties more commonly are highly variable. This variation in rock properties probably resulted from differences in initial transmissivity. Beds that had a higher transmissivity because of differences in grain size, composition, or thickness allowed greater volumes of groundwater to flow, which in turn resulted in greater dissolution of the matrix material. In contrast, beds of lower initial transmissivity remained largely unaffected by the dissolution effects of flowing groundwater. The net result of these differences can be a strongly layered reservoir.

**Fig. 2.4** is an example of dissolution layering from a core analysis of a portion of the Oligocene Frio sandstone, south Texas. Core analysis has revealed the presence of three layers of differing porosity and permeability. **Table 2.1** shows the layer properties. Dissolution caused the Type A beds to exhibit higher porosity and permeability values. The Type B sands possess a relatively high porosity but a lower permeability than the Type A beds. The lower permeability was probably caused by the larger volume of authigenic clay material deposited in the pores as products of water/mineral reactions. The micropores in the clays add to the effective porosity, but the average pore size and therefore the permeability are greatly reduced. The Type C sand has the lowest average porosity and permeability. These characteristics probably represent primary properties that were not modified by dissolution.

The lithic Frio sandstones appear to be relatively uniform and are very fine grained with a low quartz content and an abundance of volcanic-rock fragments and calcite cement. The thickness of individual beds ranges from 0.5 to 3 ft. In general, the thicker beds display better reservoir properties. Bed thickness apparently played a major role in selective dissolution and precipitation in these sandstones.

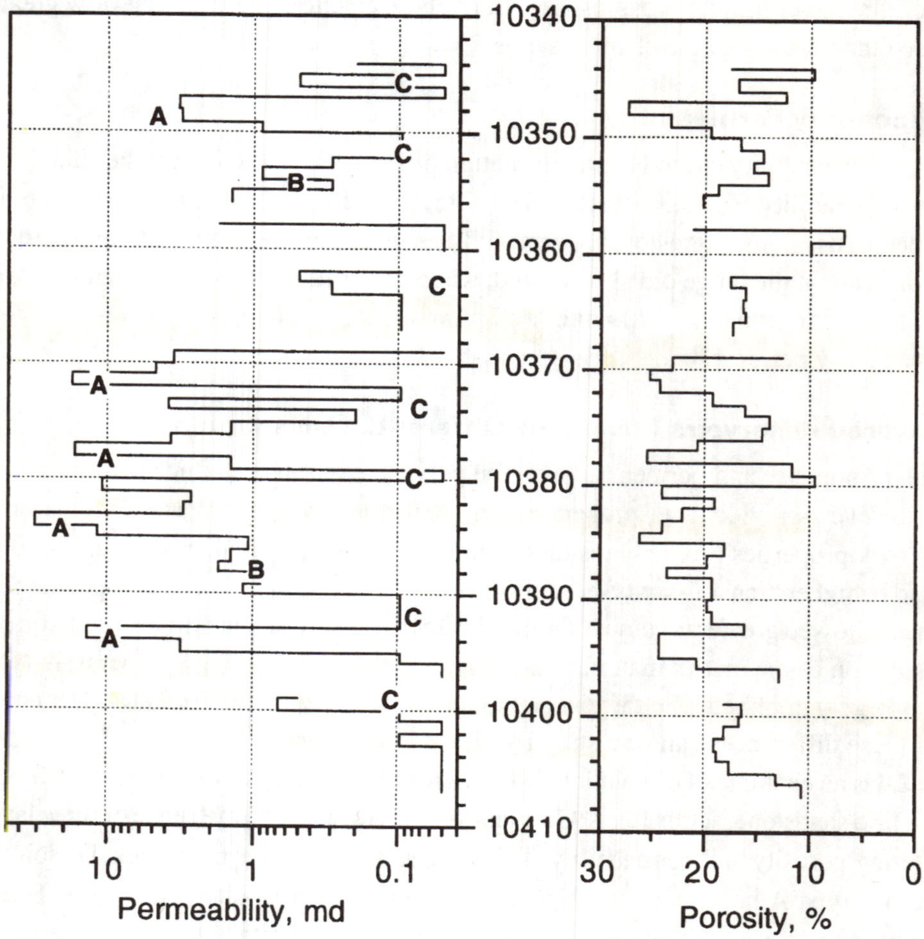


**Fig. 2.4—Core analysis of the Oligocene Frio sandstone, south Texas. The large variations in reservoir properties are the result of preferential dissolution. Types A, B, and C refer to ranges of porosity and permeability.**

## 2.5 Compressibility of Reservoir Rocks

The general compressibility equation relates the change in a unit volume of material to a change in confining pressure when the system is assumed to remain at a constant temperature.

$$c = -\frac{1}{V}\left(\frac{\partial V}{\partial p}\right)_T . \qquad (2.1)$$

| TABLE 2.1—RESERVOIR PROPERTIES, FRIO SANDSTONE, SOUTH TEXAS | | | | |
|---|---|---|---|---|
| | Permeability (md) | | Porosity (%) | |
| Porosity Type | Average | $\sigma$ | Average | $\sigma$ |
| A | 12.80 | 0.90 | 24.1 | 1.8 |
| B | 3.00 | 2.20 | 20.8 | 2.9 |
| C | 0.46 | 0.79 | 15.0 | 3.4 |

Loss of pore pressure because of production from an oil or gas reservoir causes the rock to compress in an elastic or inelastic manner. Viscous or elastic deformation implies that the compressed material will return to its original volume when the pressure differential is relieved. These effects occur in cohesive rock grains or materials that maintain their integrity during the compression/relaxation cycle.

Elastic deformation is the product of the coefficient of expansion of the material comprising the rock matrix. Deformation becomes more pronounced at high temperatures and pressures and when stress is applied over extended time periods, such as months and years. Von Gonten and Choudhary[9] measured changing compressibility values to show PV compressibility increased from 10 to 50% when the temperature was increased from 75 to 400°F. Most of the laboratory studies and correlations are related to near-room-temperature measurements. These data would not necessarily be applicable for the elevated temperature and pressure conditions normally encountered in overpressured reservoirs.

**Fig. 2.5** illustrates inelastic and elastic compression occurring during a normal burial cycle of sands and shales.[10] Shales tend to compact in an inelastic manner because the flat-lying clay platelets slide, compact, and squeeze water from between the platelets at low net overburden pressures. Gregory's[10] findings are similar to those of Geertsma.[11] These studies suggest that dewatering of adjacent undercompacted, overpressured shales acts as an energy source for mainly depletion-drive overpressured reservoirs.

Sand grains tend to bridge and form a supporting framework at low to moderate pressures. Increased burial depth usually increases the volume of cementing material, which in turn increases the strength of the matrix. The matrix will expand in an elastic manner when stress is released as long as the integrity of the system remains constant.[12] However, increasing the net overburden pressure beyond the point of matrix failure will induce microfracture formation, which is an inelastic phenomenon.

Schoonbeek[13] studied the relation between elastic and inelastic compression occurring during a burial/erosion/burial process. Burial of a poorly compacted sedimentary rock during the A→B period in **Fig. 2.6** represents a combination of elastic grain compression and inelastic crushing and meshing together of the rock material. Erosion of some of the surface deposits at Point B permit the elastic portion of the initial compression cycle to expand back to its original volume as the compression cycle migrates from B→C. Cessation of erosion and reinitiation of the burial process allows the formation to recompress

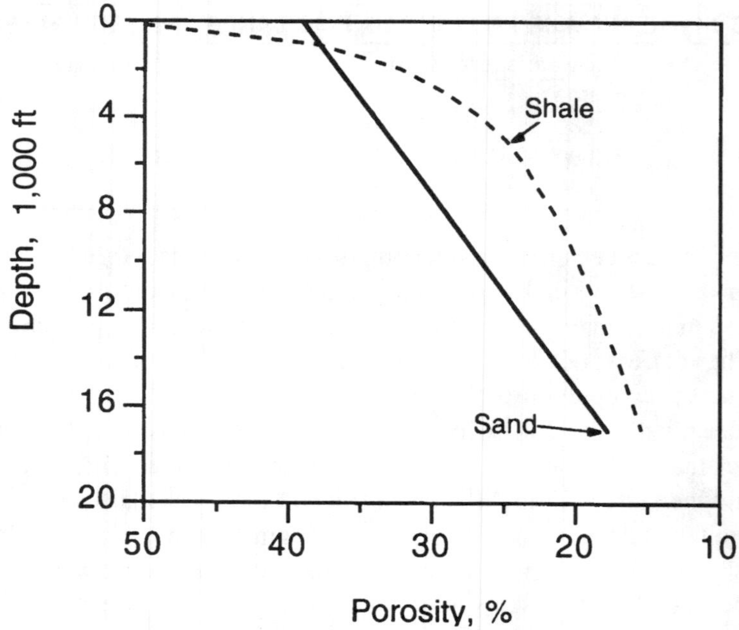


**Fig. 2.5—Relative effects of compaction on shale and sandstone porosities in a normal-pressured section (from Ref. 10).**

along the C→B line. The normal compaction process is reinstituted at Point B when the trend continues on the B→D line with increasing burial depth.

Fatt[14] uses an example of the crushing of sand grains during compaction to illustrate the inelastic/elastic hysteresis effect. **Fig. 2.7** shows the change in bulk volume of a Sespe sandstone rock sample when the pressure cycle spans two periods, 0→*p*→13,000 psi and 13,000→*p*→ 0 psi. Pore collapse at 3,500 psi altered the internal structure of the rock-sample structure. Note that the compression/expansion process is elastic in the 8,300→*p*→ 13,000-psi range because the curves retrace the same path. The divergence of the curves at lower pressures indicates that the process is inelastic.

Producing fluids from a petroleum reservoir induces a loss of supporting pore pressure. The increased stress is transferred to the rock matrix and grains. Matrix compression results when this increasing stress exceeds the strength of the porous medium. The rigidity of the matrix defines the type of inelastic deformation that may occur. Sliding and meshing together of rock grains in unconsolidated and friable material will reduce PV or formation of microfractures in a cemented rock matrix will result in a fragmented system. In either case, the effect of inelastic compression of the reservoir material is to increase the cross-sectional area and rigidity of the rock matrix. Deformation ceases when the

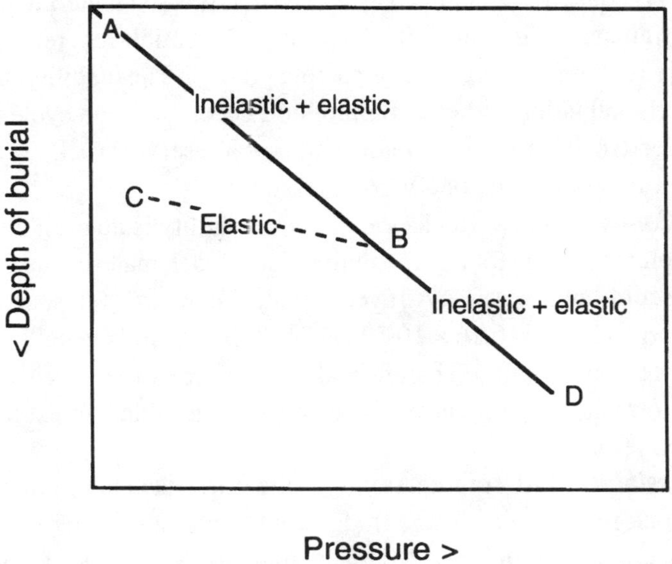


**Fig. 2.6—Effect of cyclic loading on formation compressibility (from Ref. 13).**

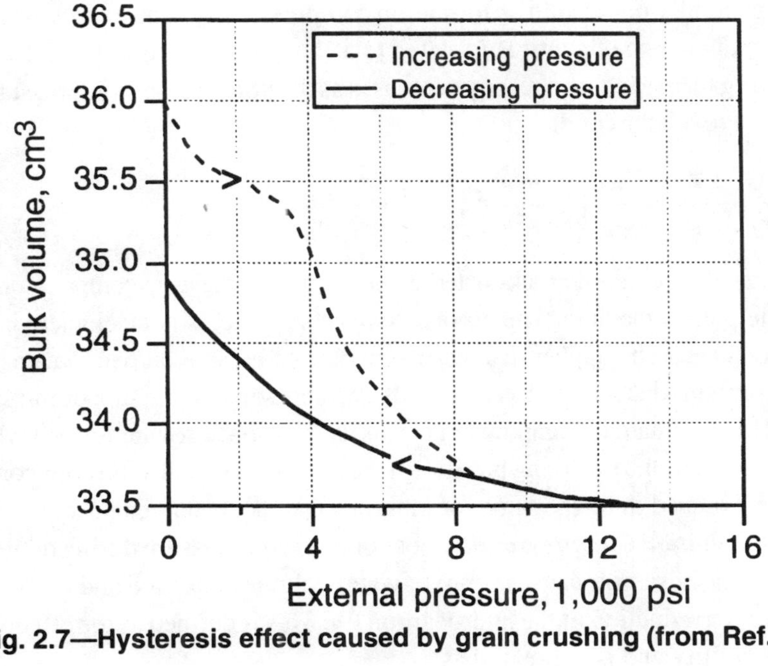


**Fig. 2.7—Hysteresis effect caused by grain crushing (from Ref. 14).**

internal strength is equal to the applied stress. The compressed material would be unable to return to its original configuration if the pressure differential were released because the internal structure is altered during the compacting process. The inability of a rock sample to return to its original volume after a compression/decompression cycle is called a hysteresis effect. Geertsma[11] states that deformation of a reservoir rock may be defined as changes in either the rock, bulk, or PV compressibility.

***Rock-Grain Compressibility.*** Rock-grain compressibility is an elastic property and is defined as the volume change for a unit volume of solid rock material measured as a function of changing confining pressure. An average value for quartz, feldspars, and rock fragments is approximately $c_{gr} = 0.16 \times 10^{-6}$ psi$^{-1}$.[15] This term is usually at least $^1/_{20}$ the formation-compressibility value. Therefore, the influence of rock-grain compressibility on reservoir performance is minimal and is usually of negligible interest when evaluating an overpressured reservoir.

***Bulk Compressibility.*** Bulk compressibility is the fractional change of a total unit bulk volume of the rock related to a change in differential pressure. Bulk compressibility is equal to the pore compressibility divided by the porosity. Subsidence studies involve the application of the bulk-volume compressibility.

***Formation Compressibility.*** Geertsma[11] defines this term as PV compressibility although it is generally referred to as formation compressibility. The expression is defined as the fractional change of a unit PV related to a change in pressure. Eq. 2.1 is rewritten in the following form to show porosity decreases as the pore or reservoir pressure declines from an initial to a later condition.

$$c_f = -\frac{1}{V_p}\frac{\Delta V_p}{\Delta p} = \frac{1}{\phi_i}\left(\frac{\phi_i - \phi_2}{p_i - p_2}\right) = \left(1 - \frac{\phi_2}{\phi_i}\right)\frac{1}{\Delta p} \quad \text{(2.2)}$$

$c_f$ is a combination of elastic rock-material expansion and inelastic compression and compaction of the porous medium. The normal compacting process is incomplete in overpressured reservoirs because a higher than normal portion of the overburden remains supported by fluid pressure. Production of reservoir fluids transfers some of the supporting stress to the underindurated rock matrix. Compression occurs as microfractures are formed when the increased stress exceeds the fracturing pressure of the rock matrix. Bulk- and pore-compressibility values are assumed to be the same for most practical petroleum engineering studies.

***Effective Formation Compressibility.*** Loss of reservoir pressure during depletion of an overpressured gas reservoir causes connate-water volume to expand and PV to compress. The net effect is a reduction of the hydrocarbon PV. This is defined as the effective formation compressibility and is expressed as

$$c_e = \frac{S_{wi} + c_f}{1 - S_{wi}}. \qquad (2.3)$$

The effective-formation-compressibility expression is a measure of the effect of changing pressure on the hydrocarbon PV. This term is particularly important when studying overpressured gas reservoirs because of the high initial pressures and usual dramatic pressure loss during depletion. A good average value for water compressibility is $c_w = 3 \times 10^{-6}$ psi$^{-1}$. PV compressibility is usually unknown and is a difficult reservoir parameter to measure because its value is a function of rock type and internal (or pore) and external (or confining) pressure stresses. These subsurface stress and rock-character conditions must be replicated in the laboratory to obtain an accurate measurement of the stress/compressibility relationship for a particular reservoir or field.

All gas reservoirs possess the same gas- and effective-formation-compressibility/pressure relationships. However, effective-formation- and gas-compressibility values are of approximately the same magnitude at elevated pressures. Therefore, both terms are included in a material-balance equation describing the performance for the overpressured reservoir case. On the other hand, the effective-formation-compressibility term is not included in a material-balance expression for the moderate- to low-pressure, depletion-drive case because the gas-expansion effect controls the performance history.

***Laboratory Measurements.*** The hydrostatic laboratory measurement method is used in the majority of published compressibility studies. The core sample is encased in an impermeable shield and inserted into a pressure chamber where internal and external pressures are maintained and measured separately. The external stress is gradually increased while the internal or pore pressure remains constant. This process compresses the liquid-saturated core. A portion of the liquid is forced from the core. The expelled liquid volume is measured as a function of confining or net overburden pressure. Eq. 2.2 converts the measured volume/pressure values to a formation-compressibility/net-overburden-pressure relationship. The hydrostatic method assumes that the forces acting along the principal axes in the subsurface are the same. This assumption is not necessarily valid when viewed in the context of the subsurface tectonic forces that may be present in a particular geologic province.

One must remember that while compressibility in the reservoir begins at some measurable net overburden stress, laboratory compressibility studies usually begin at zero net overburden stress when the pore and overburden pressures are equal. Laboratory-measured compressibilities referenced to an initial porosity should generally be greater in value than would be present in the reservoir at initial conditions.

Teeuw[16] and Lachance[17] discuss the differences between the uniaxial- and triaxial-stress compressibility measurements. Triaxial studies allow different strength stress fields to be exerted along the principal axes. Lachance found that the triaxial compressibility values are approximately one-third less than those obtained from hydrostatic measurements. These studies illustrate the difficulty of obtaining representative compressibility values for reservoirs located in tectonically active areas.

Geertsma[11] and others[15,18] concluded that deformation of consolidated sandstones is purely elastic. Shales are considered to be viscoelastic materials in which hysteresis may occur because the internal structure is easily changed during compression. Reduction of pore pressure induces inelastic compaction in poorly cemented sandstones. In reality, laboratory measurements of the PV compressibility do not differentiate between elastic and inelastic compression and a single value is usually all that is available.

**2.5.1 Reservoir-Rock-Compressibility Characteristics.** Net overburden pressure is defined as the difference between initial reservoir pressure and the current static reservoir pressure and is zero at initial conditions.

$$p_{on} = p_i - p \,. \qquad (2.4)$$

**Fig. 2.8** represents compressibility measurements obtained from the work of Fatt[15] and Fetkovich *et al.*[20] The PV compressibility values measured by Fatt increase smoothly as the net overburden pressure increases. The three-fold change in compressibility for the

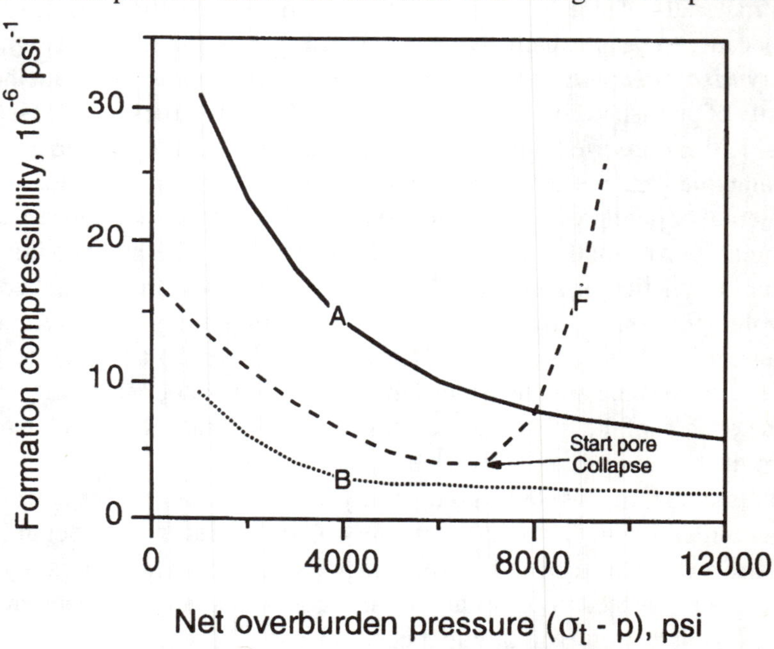


**Fig. 2.8—Effect of reservoir pressure on PV compressibility. Sample A sand grains are poorly sorted; Sample B grains are well sorted. Pore collapse began in Sample F after the net overburden pressure declined to 7,000 psi. (From Refs. 19 and 20.)**

poorly sorted Sample A containing 20 to 45% cement and intergranular detrital material is the result of compaction of the smaller-grained material into pore spaces. Bridging of the grains in the well-sorted sand in Sample B reduces the effects of compaction. Cement and detrital material comprised only 10 to 20% of the sample. Compression of a core sample increases the cross-sectional area of the supporting structure; consequently, the compressibility value decreases with increasing net overburden pressure. The Fetkovich *et al.* curve is an instance when pore collapse occurred at approximately 7,000 psia. The curve changes from smoothly decreasing to smoothly increasing at this inflection point. The Fatt samples evidently never reached the point of pore collapse.

Reservoir engineering calculations generally assume that the formation-compressibility term has a single value. Fig. 2.8 shows that the single-valued assumption is not always correct. However, the general lack of available information and our inability to quantify pressure/stress/formation-compressibility relations force us to apply this assumption to most case studies.

Harville and Hawkins[21] applied the material-balance equation to study the performance characteristics of the NS2B reservoir to calculate the formation compressibility. Their studies indicate formation compressibility varied from $c_f = 28 \times 10^{-6}$ psi$^{-1}$ at early time to $6 \times 10^{-6}$ psi$^{-1}$ at later time. The difference between Fatt's[15] laboratory studies and Harville and Hawkins' work is attributed to the fact that the rock samples are initially at equilibrium in the laboratory studies while the rock system trends toward equilibrium as pressure declines for the field case study. In other words, the endpoints are different. However, subsequent evaluation of the NS2B reservoir showed that it was also experiencing a reasonably effective waterdrive. Therefore, Harville and Hawkins' findings should be viewed with caution.

The previous discussion showed that formation compressibility is dependent on the physical characteristics of the porous medium, the variety of rock types included in the producing section, the amount and strength of the cementing material, the subsurface geologic stresses, and the degree of compaction.

**2.5.2 Effect of Net Overburden Pressure on Permeability.** Reduction of the PV by compaction and compression reduces permeability as well as porosity. Thomas and Ward[22] studied the effect of declining pore pressure on gas permeability and water saturation on a very-low-permeability, fine-grained, slightly calcareous, well-indurated sandstone. Gas permeability was reduced by approximately 80% during the first 3,000-psi decline in pore pressure although the relative permeability did not change appreciably. Initial pore pressure was set at 6,000 psi. The permeability loss was even more noticeable in fractured samples.

Vairogs *et al.*[23] studied the effect of rock stress on gas production from low-permeability gas reservoirs. **Fig. 2.9** illustrates the reduction of permeability with increased stress

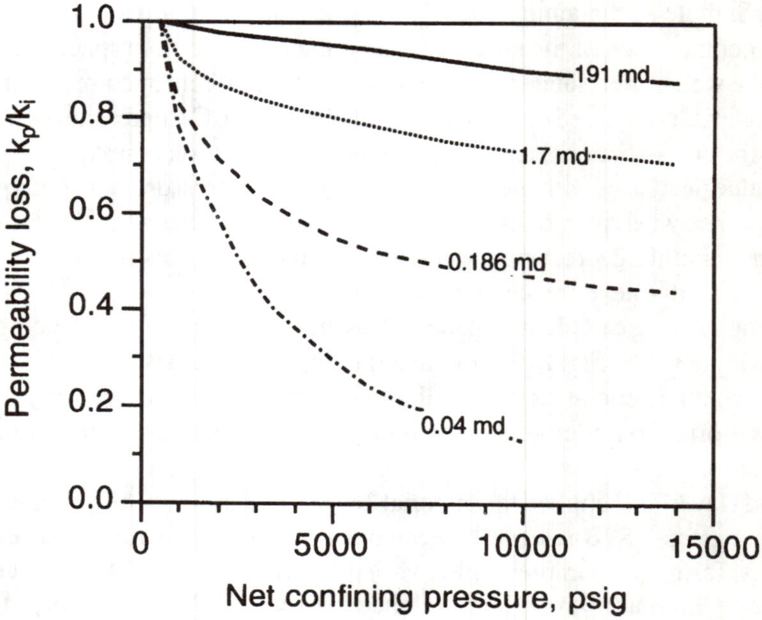


**Fig. 2.9—Effect of compaction on permeability (from Ref. 22).**

exerted on the rock sample. The permeability measurements were normalized with the permeability measured at 500 psig net pressure as the reference. The studies show that the gas-flow rate exhibited by fine-grained, overpressured reservoirs may be dramatically reduced by compaction. Better quality reservoirs are not influenced to the same degree. Formation compaction also will cause the water saturation to increase because the pore spaces will become smaller. In some cases, some of this water may become mobile.

## 2.6 Formation-Compressibility Correlations

Special-core-analysis information is seldom available for estimating formation-compressibility relations. Therefore, published correlations are widely used to determine formation compressibility for a particular reservoir study. van der Knapp's[24] experimental work indicated that pore compressibility was located somewhere between the compressibility of water, $c_w = 3 \times 10^{-6}$ psi$^{-1}$, and that of an undersaturated oil, $c_o = 1 \times 10^{-5}$ psi$^{-1}$.

The first really useful correlation was by Hall,[25] who analyzed 12 well-consolidated limestone and sandstone samples. The core samples were initially subjected to a 3,000-psi overburden pressure. The internal pore pressure varied from 0 to 1,500 psi measured at 93°F. The resulting correlation is included with Newman's[12] investigation and shown in

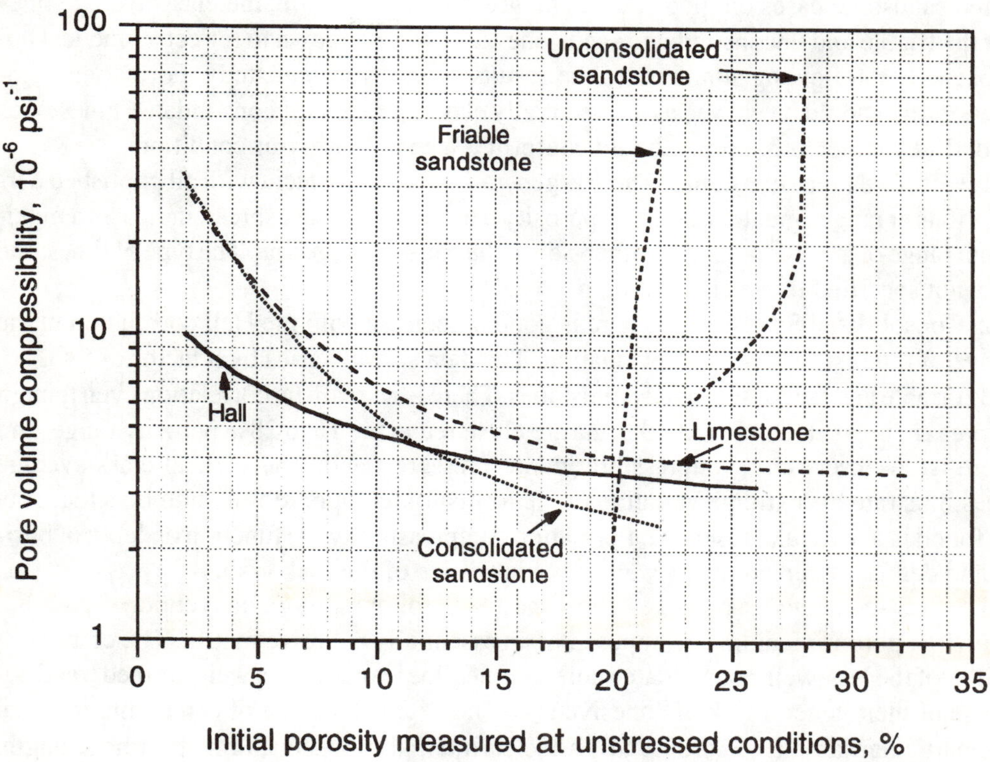


**Fig. 2.10—Comparison of the Hall and Newman compressibility estimates; Newman correlations measured at 75% of lithostatic pressure to compensate for unstressed porosity measurements (from Ref. 12).**

**Fig. 2.10**. Porosity was measured at the initial reservoir pressure when net overburden pressure is zero in the laboratory. The Hall correlation indicates that the PV compressibility is a function of volume of rock material and that the value increases as the porosity decreases. The majority of productive hydrocarbon reservoirs lie at approximately $c_f = 3.5 \times 10^{-6}$ psi$^{-1}$. The Hall correlation should be viewed with caution when studying reservoirs located in geologically young areas where the sands may be less well consolidated and should not be used for unconsolidated sands or highly fractured formations.

Newman[12] compared the results of analyzing the compressibility of 256 rock samples from 40 reservoirs with all published correlations. The core samples comprised limestone and unconsolidated, friable, and consolidated sandstone rock types. Fig. 2.10 presents these results. Class-average lines for each rock type were constructed with a statistical fitting technique. The correlating lines defined by Newman for the friable and unconsolidated sandstone cases did not fit a meaningful trend. Therefore, the class-average lines for the friable and unconsolidated sandstone cases were redrawn to reflect a practical approach. The following summarizes the results of the Newman study. Formation compressibility should be viewed as an essentially elastic process for consolidated limestones and sandstones. Correlations should conform to general reservoir conditions.

1. Consolidated Limestones. There is generally good agreement with all published correlations at ranges greater than 10% porosity. However, the data spread indicates a maximum range of a factor of five on either side of the class-average line. The type of limestone was not specified in the classification.

2. Consolidated Sandstones. There is good agreement with the Hall correlation in the range of interest for reservoir evaluation. The data spread varies by a factor of two.

3. Friable and Unconsolidated Sandstones. There is no realistic correlation even though curves are presented in Fig. 2.10. Data are clustered in the 15 to 28% porosity range, and there is no observable compressibility/pore pressure trend. The vertical class-average lines illustrate the extreme variation of the formation-compressibility numbers that may be found in a particular reservoir. Formation compressibility is a function of depth of burial and degree of cementing as well as type and size of the rock grains.

The well-cemented sandstones and limestones follow a more conventional path because the porous medium is composed of an essentially cohesive matrix. The compressibility of the less-well-consolidated core samples does not follow a well-defined trend because of their general lack of cohesiveness. The degree and type of cementing material in a particular sample will profoundly affect its internal character and supporting strength. Fig. 2.10 is particularly important because it illustrates the difficulty of using a correlation to estimate a reliable PV compressibility value for poorly consolidated sediments.

Hammerlindl[19] presented a correlation of the variation of PV compressibility with depth. The data points used in the correlation are from two different geological and geographical locations on the U.S. gulf coast. **Fig. 2.11** shows the pressure gradients of 0.7 and 0.8 psi/ft presented in the Hammerlindl correlation. Note that the entire figure is predicated on three data points obtained from south Louisiana and south Texas fields whose provenance is uncertain. However, this figure is widely accepted in the industry because of the ease of "finding a value" when little information is available to study a particular reservoir. The compressibility estimates obtained from this figure should be viewed as "qualitative" at best.

## Nomenclature

$c$ = compressibility, $Lt^2/m$, $psi^{-1}$
$c_e$ = effective compressibility, $Lt^2/m$, $psi^{-1}$
$c_f$ = formation compressibility, $Lt^2/m$, $psi^{-1}$

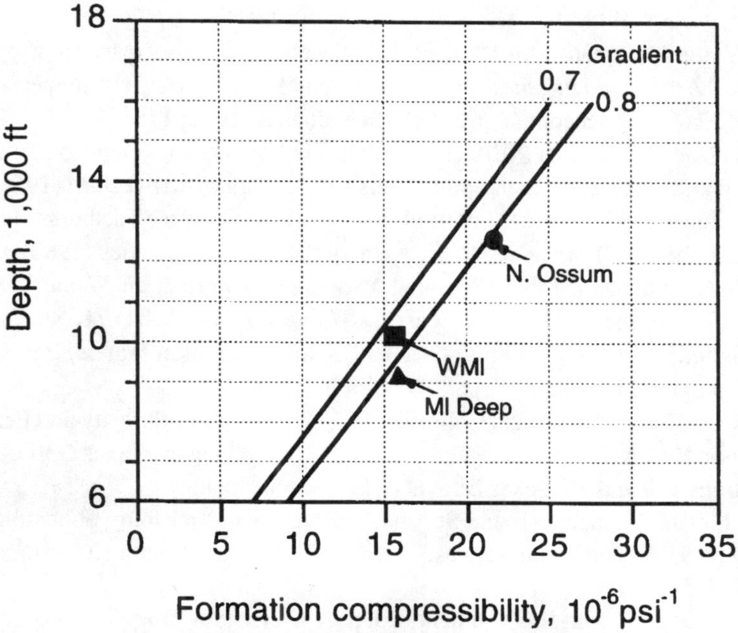


**Fig. 2.11—The Hammerlindl formation-compressibility correlation for 0.7- and 0.8-psi/ft-gradient gas reservoirs (from Ref. 19).**

$c_{gr}$ = rock-grain compressibility, Lt²/m, psi$^{-1}$
$c_o$ = oil compressibility, Lt²/m, psi$^{-1}$
$c_w$ = water compressibility, Lt²/m, psi$^{-1}$
$k_i$ = rock permeability under unstressed conditions, L², md
$k_p$ = rock permeability under stressed conditions, L², md
$p$ = pressure, m/Lt², psi
$p_i$ = initial pressure, m/Lt², psi
$p_{on}$ = net overburden pressure, m/Lt², psi
$p_2$ = pressure, m/Lt², psi
$\Delta p$ = pressure change, m/Lt², psi
$S_{wi}$ = initial water saturation
$V$ = volume, L³, cm³
$V_p$ = PV, L³, cm³
$\Delta V_p$ = PV change
$\phi_i$ = initial porosity, fraction
$\phi_2$ = porosity, fraction
$\sigma$ = standard deviation

## References


1. Schmidt, V. and McDonald, D.A.: "The Role of Secondary Porosity in the Course of Sandstone Diagenesis," *Aspects of Diagenesis*, P.A. Scholle and P.R. Schluger (eds.), Special Publication 26, Soc. of Economic Paleontologists and Mineralogists (1982) 175.
2. Atwater, G.I. and Miller, E.E.: "Effect of Decrease in Porosity with Depth on Future Development of Oil and Gas Reserves in South Louisiana," abstract, *AAPG Bull.* (1965) **49,** 334.
3. Berg, R.R. and Habeck, M.F.: "Abnormal Pressures in the Lower Vicksburg, McAllen Ranch Field, South Texas," *Trans.*, Gulf Coast Assn. of Geological Societies (1982) **32,** 247.
4. Morton, R.A. and Land, L.S.: "Regional Variations in Formation Water Chemistry, Frio Formation (Oligocene), Texas Gulf Coast," *AAPG Bull.* (Feb. 1987) **71,** No. 2, 191.
5. Land, L.S. and MacPherson, G.L.: "Origin of Saline Formation Waters, Cenozoic Section, Gulf of Mexico," *AAPG Bull.* (Sept. 1992) **76,** No. 9, 1344.
6. Loucks, R.G., Dodge, M.M., and Galloway, W.E.: "Controls of Porosity and Permeability of Hydrocarbon Reservoirs in Lower Tertiary Sandstones Along the Texas Gulf Coast," report, Investigations 149 and 78, Texas Bureau of Economic Geology.
7. Klass, M.J. *et al.*: "Diagenesis and Secondary Porosity in Vicksburg Sandstones, McAllen Ranch Field, Hidalgo County, Texas," *Trans.*, Gulf Coast Assn. of Geological Societies (1981) **31,** 115.
8. Roth, S.V.: "Depositional Environment of Upper Cretaceous Deep Tuscaloosa Sandstones, Profit Island Field, East Baton Rouge Parish, Louisiana," MS thesis, Texas A&M U., College Station, Texas (1976) 110.
9. Von Gonten, W.D. and Choudhary, B.K.: "The Effect of Pressure and Temperature on Pore Volume Compressibility," paper SPE 2526 presented at the 1969 SPE Annual Fall Meeting, Denver, CO, 28 September–1 October.
10. Gregory, A.R.: "Aspects of Rock Physics from Laboratory and Log Data that are Important to Seismic Interpretation," *Seismic Stratigraphy—Applications to Hydrocarbon Exploration*, C.E. Payton (ed.), Memoir 26, American Assn. Petroleum Geologists, Tulsa, Oklahoma (1977) 258.
11. Geertsma, J.: "The Effect of Fluid Pressure Decline on Volumetric Changes of Porous Rocks," *Trans.*, AIME (1957) **210**, 331.
12. Newman,G.H.: "Pore-Volume Compressibility of Consolidated, Friable and Unconsolidated Reservoir Rocks Under Hydrostatic Loading," *JPT* (Feb. 1973) 129.
13. Schoonbeek, J.B.: "Land Subsidence as a Result of Natural Gas Extraction in the Province of Groningen," paper SPE 5751 presented at the 1976 SPE European Spring Meeting, Amsterdam, 8–9 April.
14. Fatt, I.: "Compressibility of Sandstones at Low to Moderate Pressures," *AAPG Bull.* (Aug. 1958) 1924.
15. Fatt, I.: "Pore-Volume Compressibilities of Sandstone Reservoir Rocks" *JPT* (March 1958) 64.
16. Teeuw, D.: "Prediction of Formation Compaction from Laboratory Compressibility Data," *SPEJ* (Sept. 1971) 263; *Trans.*, AIME, **251.**

17. Lachance, D.P.: "Comparison of Uniaxial Strain and Hydrostatic Stress Pore-Volume Compressibilities in the Nugget Sandstone," paper SPE 11971 presented at the 1983 SPE Annual Technical Conference and Exhibition, San Francisco, 5–8 October.
18. Carpenter, C.B. and Spencer, G.B.: "Measurements of Compressibility of Consolidated Oil-Bearing Sandstones," Report No. RI 3540, U.S. Bureau of Mines, Washington, DC (Oct. 1940).
19. Hammerlindl, D.J.: "Predicting Gas Reserves in Abnormally Pressured Reservoirs," paper SPE 3479 presented at the 1971 SPE Fall Meeting, New Orleans, 3–5 October.
20. Fetkovich, M.J., Reese, D.E., and Whitson, C.H.: "Application of a General Material Balance for High-Pressure Gas Reservoirs," paper SPE 22921 presented at the 1991 SPE Annual Technical Meeting and Exhibition, Dallas, 6–9 October.
21. Harville, D.W. and Hawkins, M.F.: "Rock Compressibility and Failure as Reservoir Mechanisms in Geopressured Gas Reservoirs," *JPT* (Dec. 1969) 1528.
22. Thomas, R.D. and Ward, D.C.: "Effect of Overburden Pressure and Water Saturation on Gas Permeability of Tight Sandstone Cores," *JPT* (Feb. 1972) 120.
23. Vairogs, J. *et al.*: "Effect of Rock Stress on Gas Production From Low-Permeability Reservoirs," *JPT* (September 1971) 1161; *Trans.,* AIME, **251.**
24. van der Knapp, W.: "Nonlinear Behavior of Elastic Porous Media," *Trans.*, AIME (1959) **216**, 179.
25. Hall, H.N.: "Compressibility of Reservoir Rocks," *Trans.*, AIME (1953) **198**, 309.


## SI Metric Conversion Factors

| | | | |
|---|---|---|---|
| ft | × 3.048* | E − 01 | = m |
| °F | (°F − 32)/1.8 | | = °C |
| md | × 9.869 233 | E − 04 | = $\mu m^2$ |
| psi | × 6.894 757 | E + 00 | = kPa |
| $psi^{-1}$ | × 1.450 377 | E − 01 | = $kPa^{-1}$ |

*Conversion factor is exact.

## Chapter 3
# Reservoir-Fluid Properties

A natural gas is a mixture of pure components. The molecular weight, compressibility, and phase relations of each component often differ by a considerable degree. All gas-volume calculations apply a specified temperature and pressure to define the reference volume. Reference or base conditions are promulgated by local governing bodies and vary according to geography and governmental edict.

The gas-deviation or $z$ factor is a fundamental property defining the divergence of a gas volume between ideal and real conditions. The term accounts for the effect of molecular interactions on temperature/pressure/volume measurements. These nonlinear effects must be taken into account when converting gas-production volumes measured at surface conditions to reservoir conditions.

The relation of volume between reservoir and separator pressure and temperature may be obtained from a fluid-analysis study. However, these studies are seldom available and a published correlation must be used. For the most part, correlations produce acceptable results. The correlating parameters for most published studies are reduced temperature and pressure.

### 3.1 Gas-Deviation Factor

The ideal gas law relates pressure, composition, and volume.

$$pV = znRT. \qquad (3.1)$$

A correction factor is included in Eq. 3.1 to relate the difference between a gas volume measured at real and ideal conditions. This $z$ term is a function of pressure, temperature, and composition and is expressed in Eq. 3.2 as a ratio of the gas volume measured at actual and ideal conditions.

$$z = \frac{V_{\text{actual}}}{V_{\text{ideal}}}. \qquad (3.2)$$

The difference in the two volumes is nonlinear and is a function of the degree of interference between the gas molecules. All components are assumed to remain as gases over the pressure and temperature range of interest.

Natural gases are mixtures of pure hydrocarbon components that may also contain such nonhydrocarbon impurities as carbon dioxide ($CO_2$), sulfur dioxide ($SO_2$) or nitrogen ($N_2$). Each pure component gas possesses its own particular characteristics. Therefore, the $z$ factor for a multicomponent natural gas is a function of the quantity and types of pure component gases present in the mixture.

**3.1.1 Critical Properties.** The law of corresponding states assumes pure component gases possess the same fundamental characteristics when referenced to some common property.[1] The critical temperature and pressure that define the point when all liquid and vapor properties are the same are assumed to be the common references. This principle provides an technique for approximating pressure/temperature/$z$-factor relationships for hydrocarbon-gas mixtures of unknown composition.

The pseudoreduced temperature and pressure terms usually are the field parameters used to determine the $z$ factor from a correlation for a specific case study. Eqs. 3.3 and 3.4 indicate that the reduced properties are the ratios of the reservoir temperature (in degrees Rankine) and pressure (in pounds per square inch absolute) to their critical properties. The "pseudo" prefix is applied for cases when the composition of the gas mixture is unknown and the specific gravity is used to approximate the character of the gas. Pseudocritical properties are assumed to be the sum of the weighted averages of the critical properties of the individual components making up the mixture.

$$T_{pr} = \frac{T}{T_{pc}} \quad \text{(3.3)}$$

and $$p_{pr} = \frac{p}{p_{pc}}. \quad \text{(3.4)}$$

The Sutton[2] quadratic equations relate the pseudocritical pressure and temperature to the gas gravity, $\gamma_g$. Gas gravity expressed as a fraction compares the molecular weight of the gas to the molecular weight of dry air ($M = 28.97$ lbm/lbm mol). This term is generally known because it is a function of the heating value, and heating value determines the sales price of natural gas. The correlation is based on 634 compositions used in 275 laboratory studies.

$$p_{pc} = 756.8 - 131.0\gamma_g - 3.6\gamma_g^2. \quad \text{(3.5)}$$

and $$T_{pc} = 169.2 + 349.5\gamma_g - 74.0\gamma_g^2. \quad \text{(3.6)}$$

**3.1.2 $z$ Factor.** The Standing and Katz[3] chart in **Fig. 3.1** is the most widely used single-phase $z$-factor correlation for sweet hydrocarbon gases. The correlating parameters are the pseudoreduced temperature and pressure.

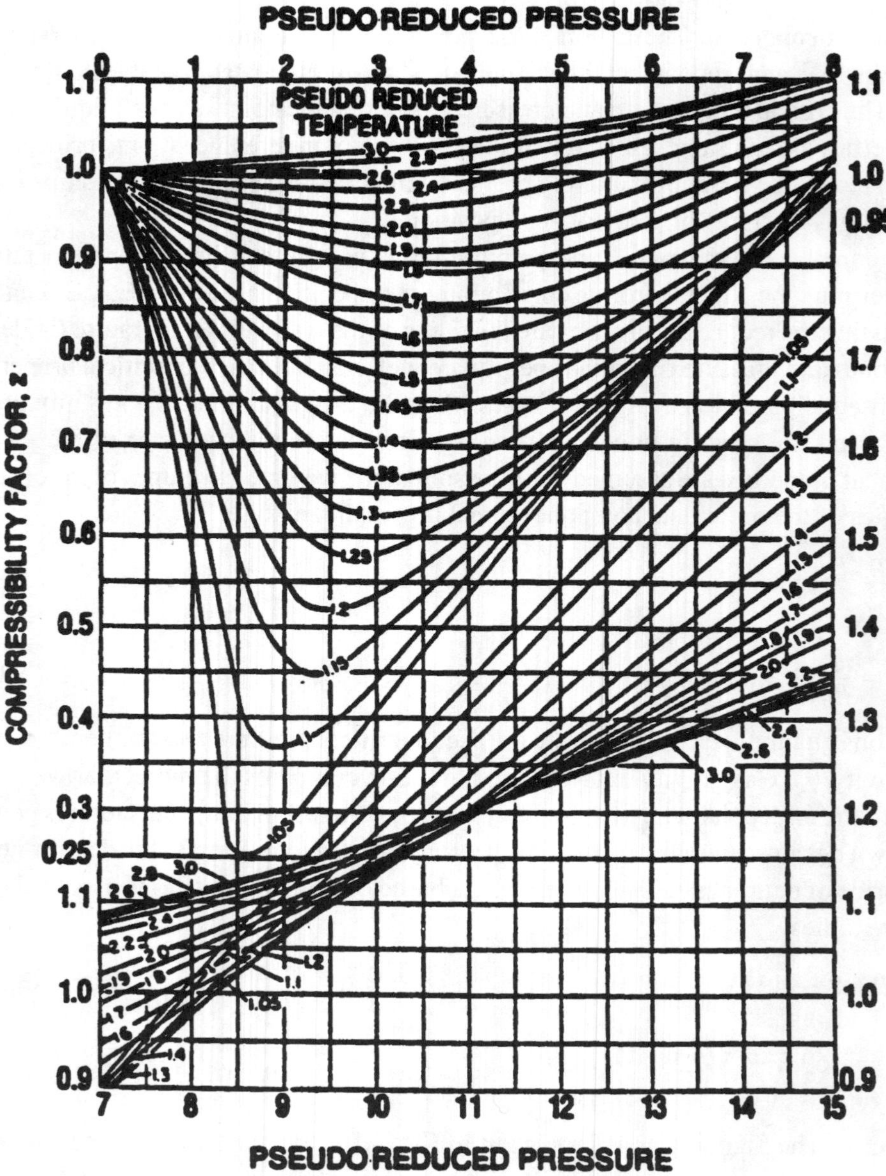


**Fig. 3.1—Standing and Katz[3] *z*-factor correlation (from Ref. 4).**

*Example Problem 3.1.* Calculate the $z$ factor for a 0.65-gravity sour natural gas when the reservoir temperature is 250°F and the reservoir pressure is 3,100 psia.

*Solution.* Use Eqs. 3.5 and 3.6 to calculate the unadjusted critical properties.

$$p_{pc} = 756.8 - 131.0\gamma_g - 3.6\gamma_g^2$$

$$= 756.8 - 131.0(0.65) - 3.6(0.65)^2 = 666 \text{ psia.}$$

$$\text{and } T_{pc} = 169.2 + 349.5\gamma_g - 74.0\gamma_g^2$$

$$= 169.2 + 349.5(0.65) - 74.0(0.65)^2 = 365^\circ\text{R.}$$

Apply Eqs. 3.3 and 3.4 to calculate the pseudoreduced temperature and pressure.

$$T_{pr} = \frac{T}{T_{pc}} = \frac{(250 + 460)}{365} = 1.95$$

$$\text{and } p_{pr} = \frac{p}{p_{pc}} = \frac{3,100}{666} = 4.66.$$

The pseudocritical temperature and pressure correlating parameters are applied to the Standing and Katz chart in Fig. 3.1 to determine the $z$ factor of the contaminated natural gas.

*Answer.* $z = 0.93$.

## 3.2 Gas Compressibility

Gas compressibility is often related to formation and connate-water compressibility in studies of overpressured gas reservoirs. The ideal gas law (Eq. 3.1) is differentiated with respect to pressure to develop a compressibility/$z$-factor relationship for constant temperature conditions.

$$\frac{dV}{dp} = \frac{nRT}{p}\frac{dz}{dp} - \frac{znRT}{p^2}. \qquad (3.7)$$

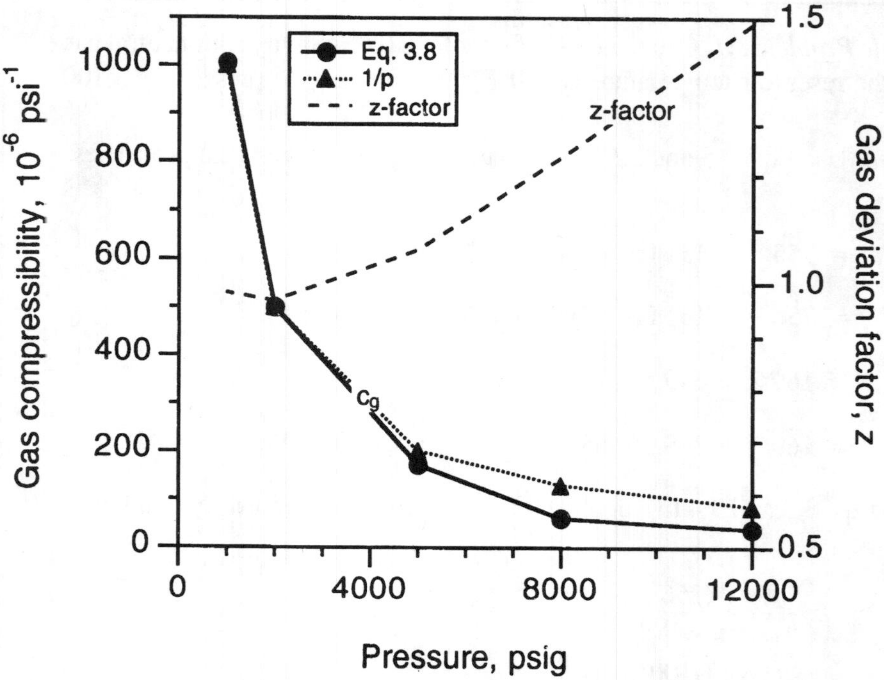


**Fig. 3.2—Comparison of gas compressibility and *z* factor as a function of pressure for a 0.7-gravity natural gas at 300°F.**

Inserting the general expression for compressibility given by Eq. 2.1 into the previous equation and simplifying results in

$$c_g = \frac{1}{p} - \frac{1}{z}\frac{dz}{dp}. \quad (3.8)$$

Eq. 3.8 is composed of two parts. Gas compressibility can be calculated with the $1/p$ portion of the equation for ideal conditions when $z = 1$ (which can occur at low to moderate pressures), a condition seldom encountered in petroleum reservoirs. However, the $z$ factor and the slope of the pressure/$z$-factor line terms expressed in Eq. 3.8 must be included when evaluating the gas compressibility at high pressures. **Fig 3.2** illustrates that gas compressibility varies by a factor of 20, while the gas-deviation term only varies by a factor of 1.5 in the 11,000-psia pressure range for the 0.7-gravity gas. Note the divergence of the $c_g$ and $1/p$ values at pressures $>5{,}000$ psig.

Pseudoreduced compressibility expresses gas compressibility in terms of pseudocritical and -reduced pressure.[5] Eq. 3.9 defines the expression. Gas compressibility values at any temperature and pressure may be directly calculated as long as the slope of the $z$-fac-

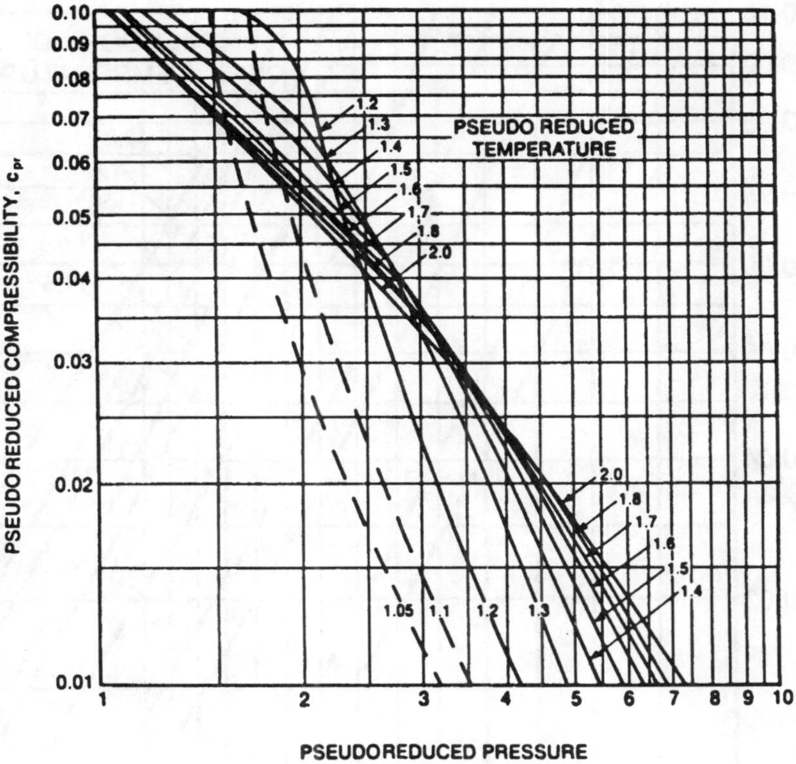


**Fig. 3.3—Trube[5] compressibility correlation for low-pressure gases (from Ref. 4).**

tor/pseudoreduced-pressure plot is known. Unfortunately, this includes an additional step in the calculations.

$$c_{pr} = \frac{1}{p_{pr}} - \frac{1}{z}\frac{dz}{dp_{pr}}. \qquad (3.9)$$

Trube[5] published a correlation relating the pseudoreduced compressibility to the pseudoreduced temperature and pressure. **Figs. 3.3 and 3.4** presents the Trube correlation for low and moderate pressures, respectively. Eq. 3.10 relates reduced to field compressibility values.

$$c_{pr} = c_g p_{pc}. \qquad (3.10)$$

Correlations do not exist for high-pressure conditions, and Eq. 3.8 must be applied to calculate gas compressibility for cases exceeding the Trube boundary conditions.

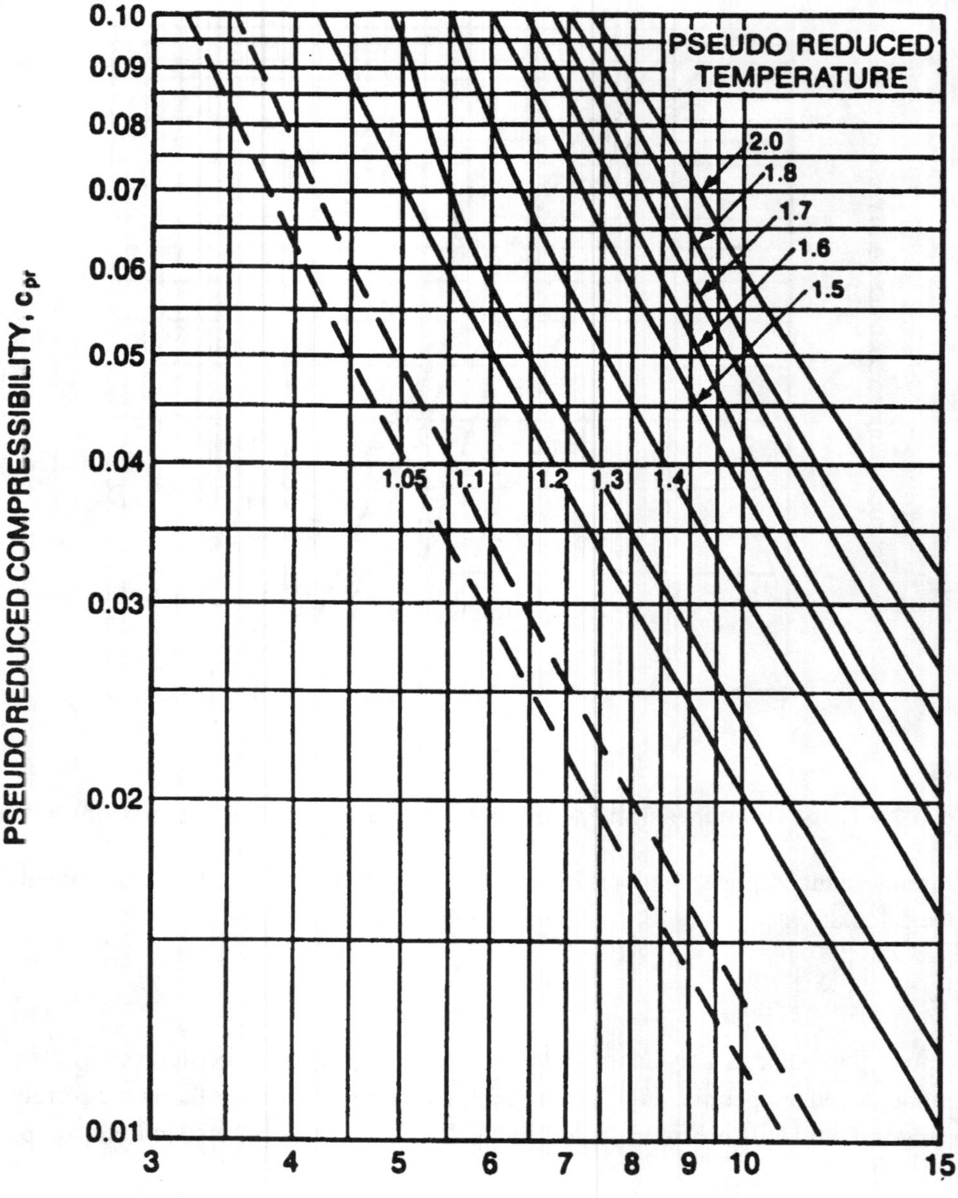


**Fig. 3.4—Trube[5] compressibility correlation for moderate-pressure gases (from Ref. 4).**

*Example Problem 3.2.* Calculate the compressibility of a 0.65-gravity gas at reservoir conditions when the temperature is 250°F and the pressure is 3,100 psia.

*Solution.* The pseudocritical and -reduced properties were calculated Example Problem 3.1 as $T_{pr}$+1.95 and $p_{pr}$+4.66. The pseudoreduced pressure and temperature are entered into Fig. 3.3 to interpret the reduced compressibility, which is $c_{pr}$+0.019. Eq. 3.10 converts the reduced compressibility value to a field value. Example Problem 3.1 calculated $T_{pc}$+365°R.

$$c_g = \frac{c_{pr}}{p_{pc}} = \frac{0.019}{666} = 2.9 \times 10^{-5} \text{ psia}^{-1}.$$

### 3.3 Gas Viscosity

Carr *et al.*'s [6] method is recommended for calculating gas viscosity. The correlation consists of two basic charts. The first chart (**Fig. 3.5**) relates temperature to molecular weight of the gas at 1 atm pressure. Inserts are included in the figure to account for the influence of $CO_2$, $N_2$, and $H_2S$ nonhydrocarbon impurities. Note that a $\gamma$ term defined as the specific gravity of the gas mixture is included in the insets to account for the presence of nonhydrocarbon gases. A correlation parameter, *K*, defined as the ratio of the gas viscosity at high pressures to the viscosity at 1 atm pressure was developed.

$$K = \mu/\mu_1 . \qquad (3.11)$$

The second figure (**Fig. 3.6**) relates pseudoreduced properties to the correlation parameter. This correlation may be used for evaluating natural gas mixtures even though it was originally developed for pure components. Fig. 3.6 relates the viscosity determined in Fig. 3.5 at 1 atm to viscosity at reservoir pressure. The correlation variables are $\mu/\mu_1$ vs. pseudoreduced pressure.

Eq. 3.12 relates the molecular weight of a gas to the specific gravity.

$$M_g = 28.96\gamma_g . \qquad (3.12)$$

*Example Problem 3.3.* Determine the viscosity of a sour natural gas at a temperature of 250°F and a pressure of 3,100 psia. The percentage impurities in the 0.65-gravity gas are 2% $N_2$, 2% $CO_2$, and 14% $H_2S$.

*Solution.* Use Eq. 3.12 to calculate the molecular weight.

$$M_g = 28.96\gamma_g = 28.96(0.65) = 18.8 \text{ lbm mol.}$$

Find $m_g$ at 1 atm. The first part of the correlation is shown in Fig. 3.5. Apply the gas properties $M$+18.8 lbm/lbm mol and $T$+250°F values to find the gas viscosity at 1 atm.

$$\mu_1 = 0.0132 \text{ cp.}$$

Adjust for the effect of impurities. From the inserts shown in Fig. 3.5, the correction for 2% $N_2$ is +0.0002 cp, the correction for 2% $CO_2$ is +0.0001 cp, and the correction for 14% $H_2S$ is +0.0003 cp for a total correction of +0.0006 cp.

$$\mu_1 = 0.0132 + 0.0006 = 0.0138.$$

For all practical purposes, the impurities do not materially affect the viscosity.

Adjust the gas viscosity estimated at 1 atm to reflect reservoir conditions. Pseudoreduced properties were calculated in Example Problem 3.1 as $T_{pr}$+1.95 and $p_{pr}$+4.66. From Fig. 3.6, $m/m_1$+1.4. Applying Eq. 3.11,

$$\mu_g = 1.4(0.0138) = 0.0193 \text{ cp.}$$

## Nomenclature

$c_g$ = gas compressibility, $Lt^2/m$, $psi^{-1}$
$c_{pc}$ = pseudocritical compressibility, $Lt^2/m$, $psi^{-1}$
$c_{pr}$ = reduced compressibility, $Lt^2/m$, $psi^{-1}$
$K$ = correlation term
$M$ = molecular weight, m, lbm/lbm mol
$M_g$ = gas molecular weight, m, lbm/lbm mol
$n$ = number of moles
$p$ = pressure, $m/Lt^2$, psi

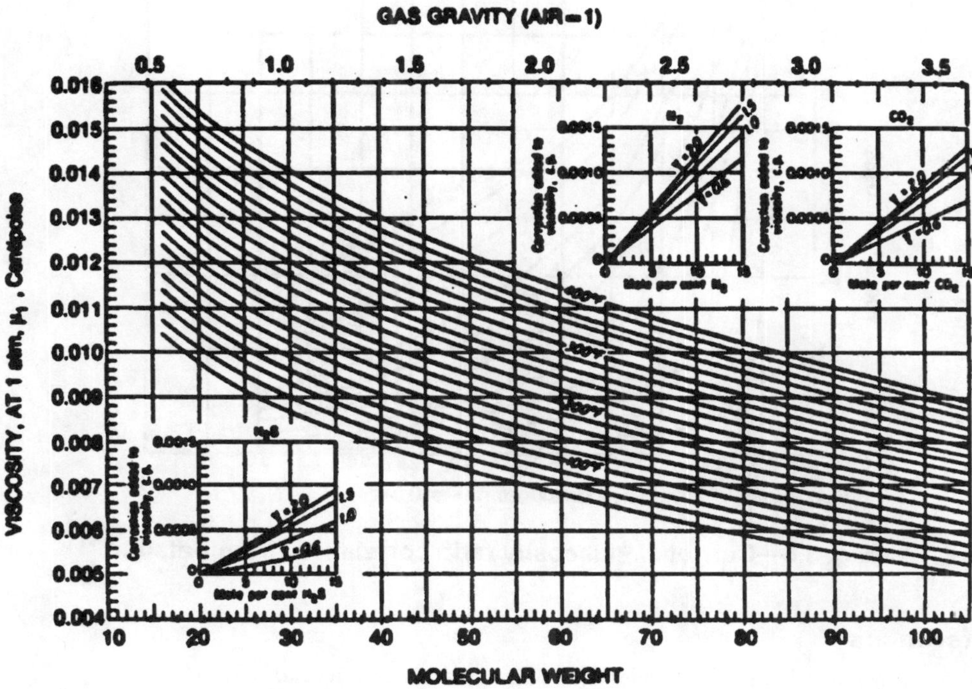


**Fig. 3.5—Carr *et al.*[6] viscosity of gases correlation. The *g* term accounts for the effect of the gas specific gravity in the impurity inserts (from Ref. 2).**

$p_{pc}$ = pseudocritical pressure, m/Lt$^2$, psi
$p_{pr}$ = pseudoreduced pressure, m/Lt$^2$, psi
$R$ = universal gas constant, mL$^2$/t$^2$T
$T$ = temperature, T, °R
$T_{pc}$ = pseudocritical temperature, T, °R
$T_{pr}$ = pseudoreduced temperature, T, °R
$V$ = volume
$V_{\text{actual}}$ = actual volume, L$^3$
$V_{\text{ideal}}$ = ideal volume, L$^3$
$z$ = gas-deviation factor
$\gamma_g$ = gas specific gravity
$\mu$ = viscosity, m/Lt, cp
$\mu_g$ = gas viscosity, m/Lt, cp
$\mu_1$ = viscosity at 1 atm, m/Lt, cp

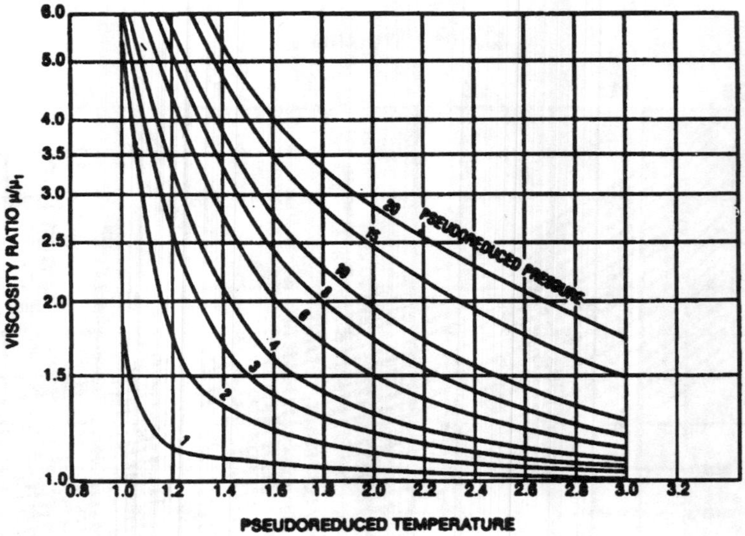


**Fig. 3.6—Carr *et al.*[6] viscosity ratio correlation (from Ref. 4).**

## References


1. van der Waals, J.D.: *Proc.*, Academy of Science, Amsterdam (1901) **3,** 515.
2. Sutton, R.P.: "Pseudocritical Properties of Natural Gases," paper SPE 14265 presented at the 1985 Annual Technical Conference and Exhibition, Las Vegas, NV, 22–25 September.
3. Standing, M.B. and Katz, D.L.: "Density of Natural Gases," *Trans.*, AIME (1942) **146,** 140.
4. Metcalf, R.S.: "Gas Properties and Correlations" *Petroleum Engineering Handbook,* H.R. Bradley (ed.), SPE, Richardson, TX (1987) Chap. 20.
5. Trube, A.S.: "Compressibility of Natural Gases," *Trans.*, AIME (1957) **210,** 355.
6. Carr, H.L., Kobayashi, R., and Burrows, D.B.: "Viscosity of Hydrocarbon Gases Under Pressure," *Trans.*, AIME (1954) **201,** 23.


## SI Metric Conversion Factors

| | | | |
|---|---|---|---|
| cp | × 1.0* | E − 03 | = Pa·s |
| ft | × 3.048* | E − 01 | = m |
| °F | (°F − 32)/1.8 | | = °C |
| lbm | × 4.535 924 | E − 01 | = kg |
| lbm mol | × 4.535 924 | E − 01 | = kmol |
| psi | × 6.894 757 | E + 00 | = kPa |
| $psi^{-1}$ | × 1.450 377 | E − 01 | = $kPa^{-1}$ |
| °R | 5/9 | | = K |

*Conversion factor is exact.

## Chapter 4
# Reservoir Mechanics

The primary influence on performance history of depletion-drive gas reservoirs is the value of the gas-compressibility term. However, water influx must be included with the gas-compressibility effect when combination or complete waterdrive reservoirs are studied. Overpressured gas reservoir performance histories also may be affected by three additional but subordinate drive mechanisms. These potential energies are effective formation compressibility, which is a function of rock-grain and connate-water expansion and formation compaction; possible influx of water from adjacent shales; and liberation of solution gas from connate and aquifer water. Theoretically, each of these energy potentials must be identified and their effects quantified to predict reservoir performance. In reality, this is a virtual impossibility, and these subordinate categories are often included as a basket of uncertain values in the effective-formation-compressibility expression. One would expect the influence of these subordinate drives to be of most interest at early times because gas compressibility is at its lowest value at initial conditions. The increased size of the gas-compressibility term at moderate to low pressures usually nullifies the influence of subordinate drive mechanisms.

### 4.1 General Basic Material-Balance Equation

**Fig. 4.1** shows initial and later conditions for the case of an overpressured, partial-water-drive gas reservoir. Gas and water production depleted the reservoir, and the pressure has declined from an initial to some reduced value. All volumes are measured at standard conditions to ensure a uniform basis of expression. The left side of the figure represents the initial reservoir-gas volume measured at initial conditions. The component volumes shown on the right side of the figure are evaluated at a lower reservoir pressure and are listed in their approximate order of importance.

Production of reservoir fluids reduces the pore pressure. The net effect of the pressure loss is a reduction in the hydrocarbon pore volume (HCPV) because of the expansion of the connate water and rock material. Water influx may come from an aquifer or from adjacent undercompacted shales. Solution gas may be liberated from the connate and aquifer water if a significant decline in reservoir pressure occurs. Reservoir volume at later conditions is expressed as the sum of four factors: expansion of the remaining gas; net aquifer water influx; change in HCPV caused by the effective-formation-compressibility effects; and the basket of the shale-water, liberated-gas pressure-dependent, and time-dependent subordinate energy terms. A material-balance equation can mathematically account for

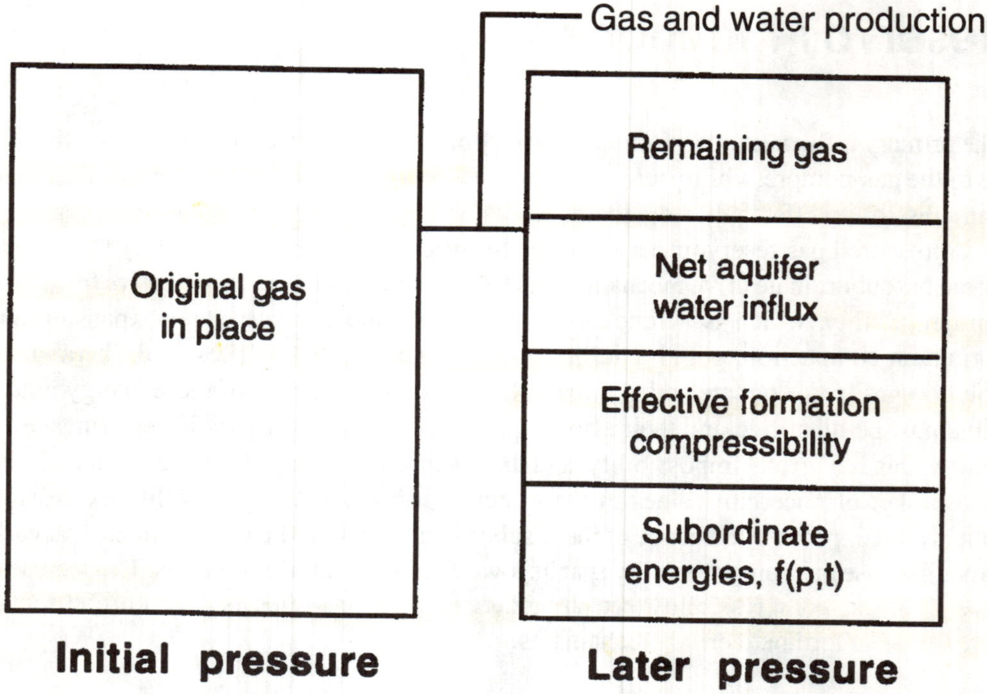


**Fig. 4.1—Effects of gas and water production on an overpressured gas reservoir.**

all these major and minor potential energies in a single expression. Each category is developed in the following discussion.

**4.1.1 Gas Expansion**. The initial-gas volume is subtracted from the produced-gas volume to calculate the volume of gas remaining at the later conditions. These volumes are evaluated at a common reservoir pressure and are related to base conditions in units of reservoir barrels per standard cubic foot. Reservoir volumes for all components of the material-balance equation are maintained in barrels for this formula.

The expansive effects of the gas remaining in the reservoir as pressure declines from $p_i \rightarrow p$ provides an expanding effect to lessen the pressure loss caused by reservoir voidage. The remaining gas volume in terms of the $z$ factor is

$$G_r = 0.00504\left(G - G_p\right)\frac{zT}{p}. \quad (4.1)$$

**4.1.2 Net Water Influx.** Net water influx is defined as the difference in the total volume of water that has encroached into the reservoir minus the volume of produced water. The term is usually assumed to represent only aquifer-water influx. The effect of shale-water encroachment is generally not included even though the effects of the two are virtually indistinguishable.

$$W_{en} = \left(W_e - W_p\right)B_w . \quad (4.2)$$

**4.1.3 Effective Formation Compressibility.** Eq. 2.3 indicated that effective formation compressibility is a function of formation- and connate-water compressibility and HCPV. Additionally, the discussion in Chap. 2 showed that this term is variable and should not be treated as a single-valued function. An equation reflecting the changes in the HCPV caused by declining reservoir pressure is developed by combining the general expression for compressibility (Eq. 2.2) with definitions for connate water and pore volumes (PV's).

$$\Delta V_w = - c_w V_w \Delta p = 0.00504 c_w \left(p_i - p\right)\left(\frac{GS_{wi}}{1 - S_{wi}}\right)\left(\frac{Tz_i}{p_i}\right). \quad (4.3)$$

The decrease in PV caused by the incremental pressure decline is equated in the same manner.

$$\Delta V_p = - c_f V_p \Delta p = 0.00504 c_f \left(p_i - p\right)\left(\frac{G}{1 - S_{wi}}\right)\left(\frac{Tz_i}{p_i}\right). \quad (4.4)$$

The total loss in HCPV caused by the increased connate-water volume and decreased PV is

$$\Delta V_w + \Delta V_p = 0.00504 \frac{G\left(p_i - p\right)}{1 - S_{wi}}\left(c_w S_{wi} + c_f\right)\left(\frac{Tz_i}{p_i}\right). \quad (4.5)$$

**4.1.4 Pressure- and Time-Dependent Potential Function.** The basket of pressure- and time-dependent shale-water influx, $W_{e(sh)}$; connate-water solution-gas/-oil ratio, $R_{s(cw)}$; and aquifer-water solution-gas/-oil ratio, $R_{s(aq)}$ subordinate energy terms are included in the pressure- and time-dependent basket of terms, the $f(p,t)$ expression. These terms are extremely difficult to quantify and are often included in the effective-formation-compressibility expression simply because of their vagueness and uncertain effects on reservoir performance history for most case history studies.

Summing and rearranging the four major energy categories gives

$$\frac{p}{z}\left[1 - \frac{198.4\left(W_e - W_p\right)B_w p_i}{\left(p_i - p\right)Gz_i T} - \left(\frac{c_w S_{wi} + c_f}{1 - S_{wi}}\right)\left(p_i - p\right) - f(p,t)\right]$$

$$= \frac{p_i}{z_i} - \frac{p_i G_p}{z_i G}. \quad (4.6)$$

Application of the following reservoir energy definitions shortens Eq. 4.6 to a more readable expression. Net water influx is

$$W_{en} = \frac{198.4(W_e - W_p)B_w p_i}{(p_i - p)G z_i T}; \quad (4.7)$$

from Eq. 2.3, effective formation compressibility is

$$c_e = \frac{c_w S_{wi} + c_f}{1 - S_{wi}};$$

and the pressure-/time-dependent potential function is

$$\mathrm{f}(p,t) = f\left[W_{e(sh)} + R_{s(cw)} + R_{s(aq)}\right]. \quad (4.8)$$

Eq. 4.6 can now be expressed in a component form as

$$\frac{p}{z}\left[1 - W_{en} - c_e - \mathrm{f}(p,t)\right] = \frac{p_i}{z_i} - \frac{p_i G_p}{z_i G}. \quad (4.9)$$

The $\mathrm{f}(p,t)$ term is included in Eq. 4.9 for completeness. However, it is usually ignored because of the general uncertainties of quantifying these usually minor energy effects.

Some combination of these components listed in the material-balance equation influences the performance history of all gas reservoirs. The following discusses these influences in the context of pressure-depletion and combination pressure-depletion/water-drive gas reservoirs.

## 4.2 Pressure-Depletion Gas Reservoirs

The gas-expansion effect is the primary drive mechanism for normal-pressured depletion gas reservoirs. These reservoirs can be divided into two types: (1) the over- or very-high-pressure case where the gas-, formation-, and connate-water-compressibility values are of similar magnitude at initial conditions but diverge with declining pressure and (2) the normal-pressured case where the gas effect is the predominant drive mechanism.

**Fig. 4.2** illustrates the effect of changing pressure on gas-, formation-, and water-compressibility values for the Anderson L reservoir. Gas compressibility[1] increased from $26 \times 10^{-6}$ to $232 \times 10^{-6}$ psi$^{-1}$, while water[1] and formation compressibility[2] remained essentially constant at $c_w \approx 3 \times 10^{-6}$ psi$^{-1}$ and $c_f \approx 15 \times 10^{-6}$ psi$^{-1}$, respectively, over a 6,000-psi pressure range. The figure indicates that water and formation compressibility

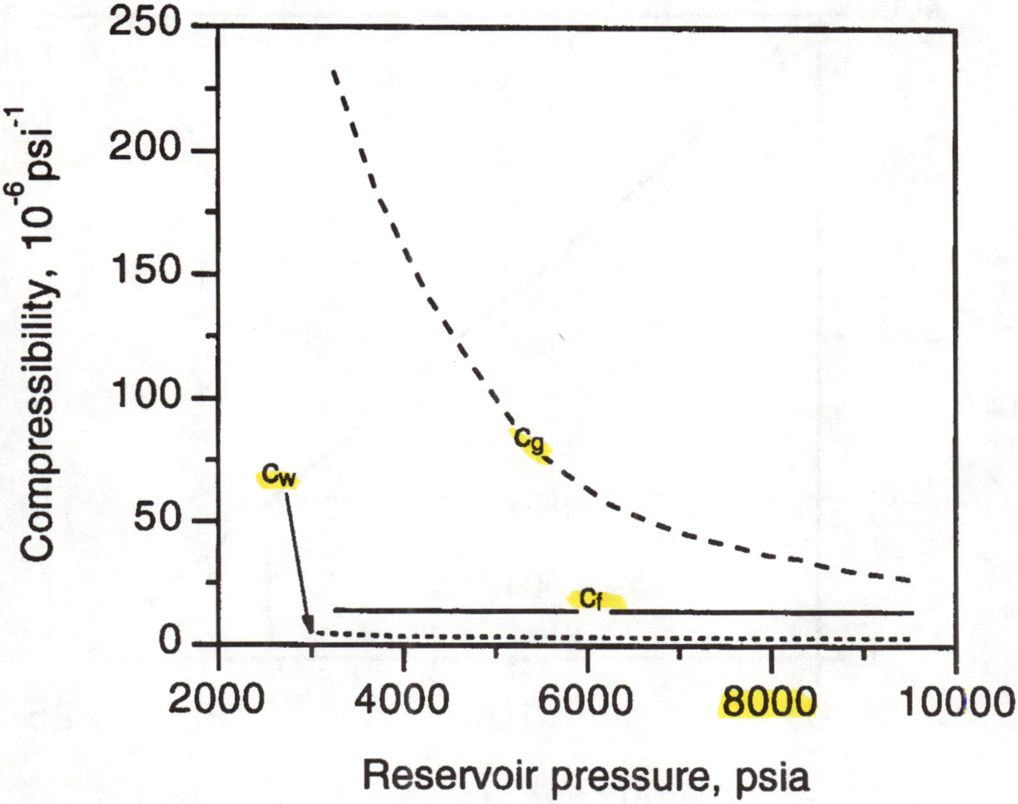


**Fig. 4.2—Effect of pressure on gas, water, and formation compressibility from the Anderson L reservoir [from Refs. 1 (gas and water compressibility) and 2 (formation compressibility)].**

will probably play a major role in reservoir performance history only at pressures > 8,000 psia. The large value of the gas-compressibility term essentially nullifies the potential for the effective-formation-compressibility term to alter performance behavior at moderate to low reservoir pressures.

**4.2.1 Moderate- to Low-Pressure Reservoirs.** Eq. 4.9 is shortened to the following form when water influx and the subordinate energy terms are assumed to be absent and the initial reservoir pressure is less than ≈6,000 psia.

$$\frac{p}{z} = \frac{p_i}{z_i} - \frac{p_i}{z_i}\frac{G_p}{G}. \qquad (4.10)$$

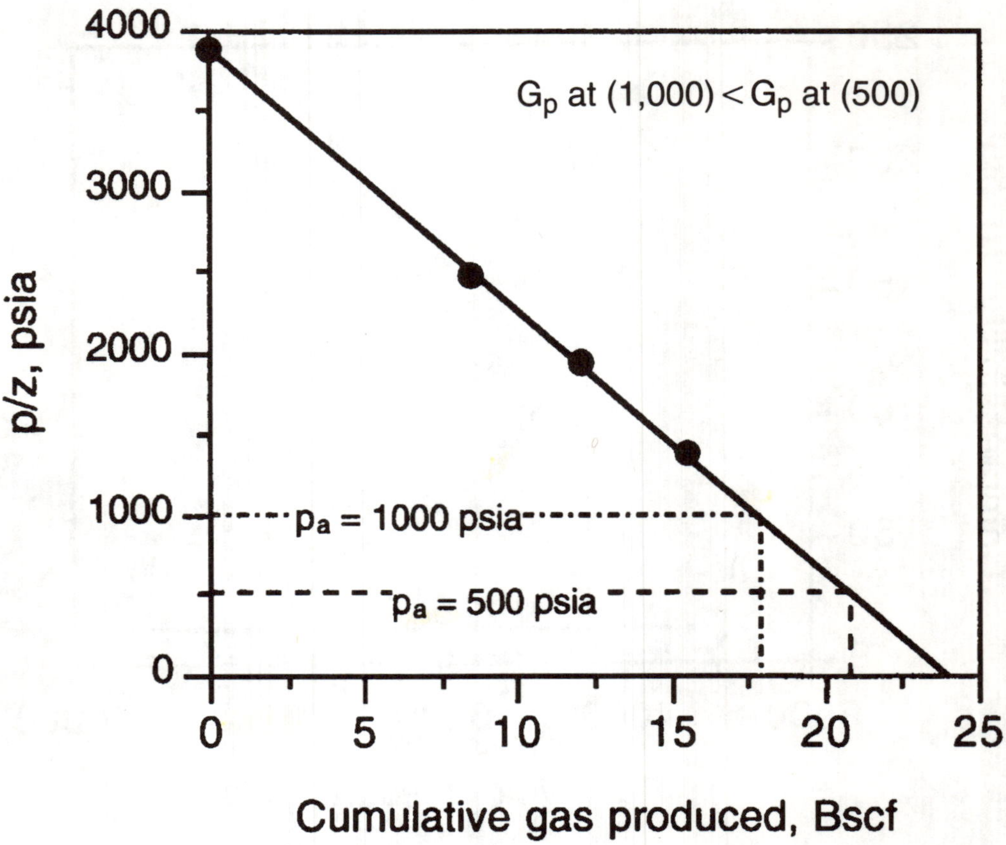


**Fig. 4.3—Example *p/z* plot for a pressure-depleting reservoir.**

Gas compressibility is assumed to supply all the energy for producing normal-pressured depletion-type gas reservoirs. A straight-line relationship results when the $p/z$ term is plotted on the ordinate and $G_p$ is plotted on the abscissa. The endpoint of the straight line is an estimate of the volume of gas originally in place because pressure at this point is zero. **Fig. 4.3** illustrates the straight-line $p/z$-$G_p$ plot, more commonly referred to as a $p/z$ plot. Ultimate recovery is directly related to abandonment pressure. The figure shows that installing a compressor would allow the abandonment pressure to be lowered from 1,000 to 500 psia and increase reserves by approximately 2.9 Bscf. The effect of decreasing the flowing bottomhole pressure on ultimate recovery is easily visualized. The $p/z$ plot will remain a straight line as long as the reservoir depletion drive is not affected by water influx or some other alteration of the reservoir system.

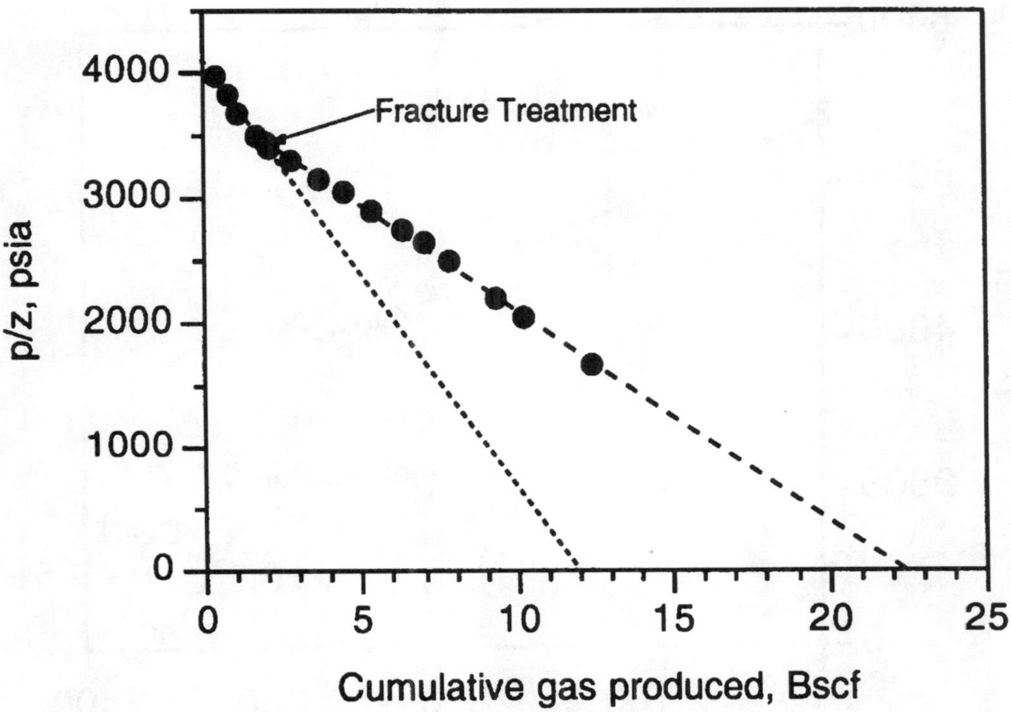


**Fig. 4.4—Effect of increasing drainage radius of Well 2 on the *p/z* plot (from Ref. 13).**

Stewart[3] discussed a number of field examples where the slope of the *p/z* line of a well producing from a pressure-depleting reservoir will be affected when the drainage area is changed. **Fig. 4.4** shows the ultimate recovery in Well 2 increased from 12.0 to 22.4 Bscf when a hydraulic fracture treatment extended the drainage radius. Two straight lines are evident. The share of the total field production of Well 2 increased from 4.2 to 8.8 %.

**4.2.2 Overpressured Reservoirs. Fig. 4.5** is a typical example of the concave or dual-slope nature of a *p/z* plot of an overpressured gas reservoir. The flat-lying slope of the early-time curve indicates that an energy source other than gas compressibility must also be influencing performance history.[4] Fig. 4.2 illustrates the variable compressibility effects causing this duality. Rock- and water-compressibility values approach the gas-compressibility value at high pressures. Therefore, these two terms must be included in a material-balance equation simulating the performance of overpressured gas reservoirs, at least at early-time conditions. The standard *p/z* plot extrapolates to the true original gas

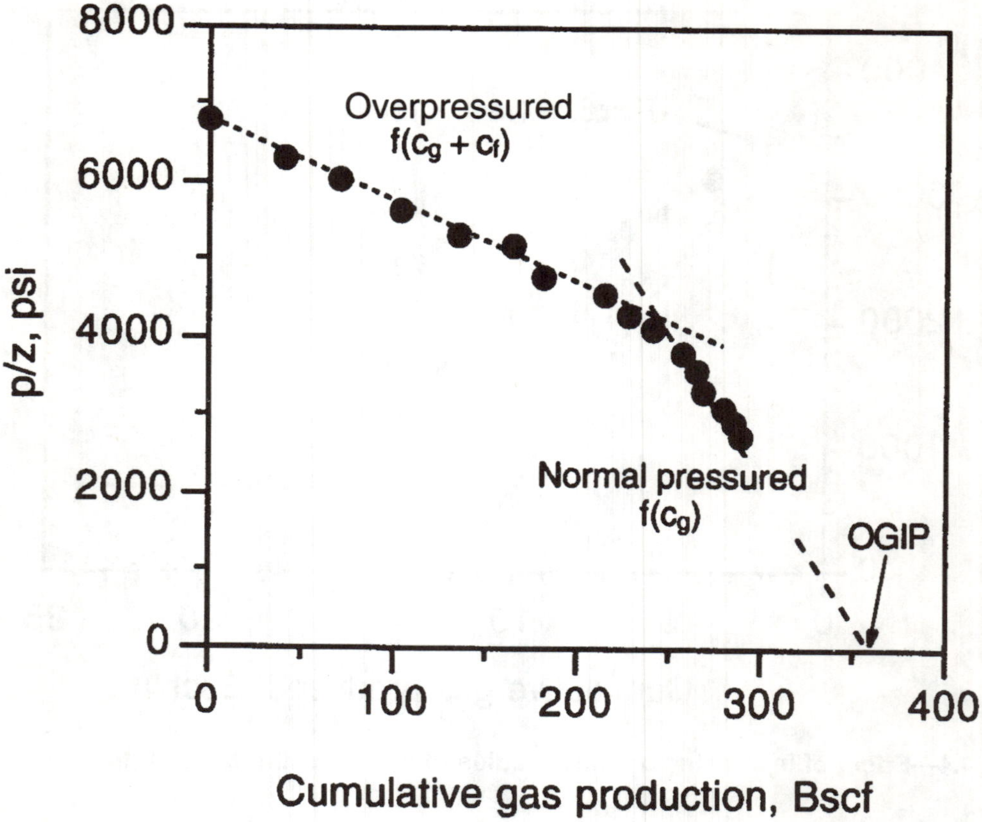


**Fig. 4.5—A typical dual-slope *p/z* plot for an overpressured gas reservoir.**

in place (OGIP) at moderate to low pressures because gas compressibility controls the performance history over this pressure range. Ramagost and Farshad[5] showed that inclusion of the effective-formation-compressibility term in the material-balance solution for the overpressured depletion-drive gas reservoir case resulted in a straight line.

Production of reservoir fluids reduces pore pressure. This causes connate-water and rock volumes to expand. Compaction may also occur if the stress exceeds the strength of the rock matrix. The effects of expanding connate water and the formation are included in the effective-compressibility term (Eq. 2.3). The work of Fetkovich *et al.*[1] discussed in Sec. 2.5.1 showed that formation compressibility may be a dual-slope function. The value can either decrease if the reservoir rock is in the compression stage or increase if the reservoir rock is in the failure stage. The reservoir-pressure range defines where a particular reservoir is operating on the compressibility curve. Unfortunately, the true stress/time relationships exerted in a reservoir can seldom be determined.

The effect of the changing PV-compressibility term is most pronounced in overpressured gas reservoirs because the usual geological compaction and consolidation processes have not been completed. Therefore, average porosity is greater than and the strength of the rock matrix is less than would be found in comparable normal-pressured sections. Hammerlindl[2] states that compaction is essentially complete when the reservoir pressure has declined to a normal gradient. PV compressibility will become of negligible importance after this point because the value of the gas-compressibility term is many times greater than the formation-compressibility value.

Refs. 1, 6, and 7 include the effect of liberated solution gas in the reservoir water as a possible energy source when studying overpressured gas reservoirs. Fetkovich *et al.*[1] showed that the initial solution-gas/-water ratio for the Anderson L reservoir of 31.8 scf/STB at 9,507 psia declined to 11.8 scf/STB when the reservoir pressure was 3,000 psi. However, the reservoir pressure must decline to approximately 6,700 psia before appreciable volumes of solution gas are liberated. The gas-compressibility term would have increased at this point to a value that would mask these solution-gas-drive effects in most cases. Additionally, the presence of a large volume of water to increase the solution-gas effect would imply the presence of a large aquifer, which in turn would imply increased water-influx effects. The reservoir energy from water influx would probably nullify the importance of the liberated-solution-gas term in this case. Liberated solution gas will be an important source of reservoir energy only in certain cases.

## 4.3 Waterdrive Gas Reservoirs

The effect of water influx adds an additional displacement process and pressure-supplement source to the reservoir. Residual-gas saturation is reduced but abandonment pressure is increased in the presence of water influx. Generally, the net effect is to reduce ultimate recovery. Formation compressibility for most normal-pressured aquifers is assumed to range from $3 \times 10^{-6}$ to $4 \times 10^{-6}$ psi$^{-1}$. Undercompacted, overpressured aquifers should display a somewhat larger value. The effect of increasing the formation compressibility would be to decrease the apparent size of the aquifer.

**4.3.1 Moderate- to Low-Pressure Reservoirs.** Bruns *et al.*[8] conducted a study of the effect of varying aquifer transmissibility and size on the $p/z$ plot. Water influx caused the theoretical straight line of the $p/z$ plot to shift to the right and become curved with at least one and sometimes two inflection points. **Fig. 4.6** is an example of their work. Aquifer size is defined as the ratio of the radius of the aquifer compared with the radius of the reservoir, $r_D = r_{aq}/r_o$, where the subscripts $aq$ and $o$ define the radius of the aquifer and gas reservoirs, respectively. The curves displayed an essentially similar curvature for all val-

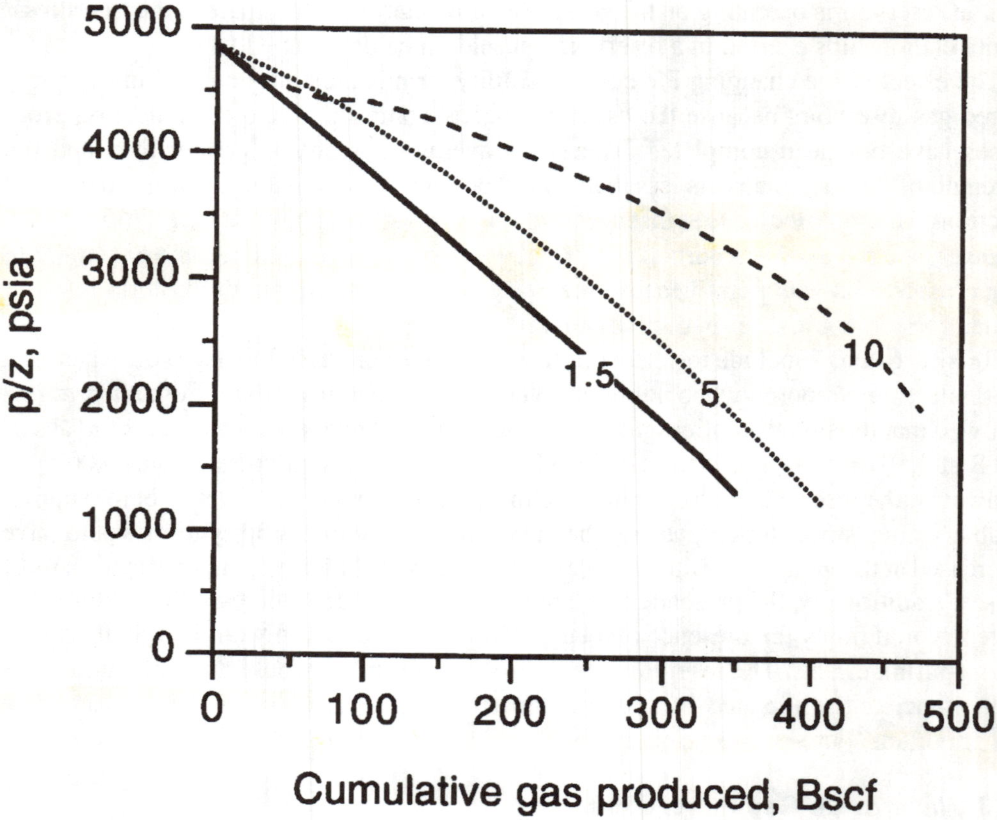


**Fig. 4.6—The effect of increasing $r_D$ values on the *p/z* plot (from Ref. 8).**

ues of $r_D \leqq 5$. However, aquifers with a permeability of $k < 100$ md generally showed an additional inflection point at early time. This additional inflection point is caused by the amount of time required before the aquifer is able to furnish appreciable water influx to augment reservoir pressure. The presence of a high permeability and a large aquifer allows the repressuring effect of the encroaching water to be quite pronounced even at early time. In any case, the effect of water influx is to shift all curves to the right.

The $p/z$ plot of the $r_e/r_o = 1.5$ aquifer size is only slightly curved in the 300- to 400-Bscf region. Pressure support from the aquifer is of minor importance in this instance when essentially total-pressure-depletion conditions exist. The effects of water influx may be minimized if reservoir production is maintained at a sufficiently high rate. This is the reason for depleting gas reservoirs at as high a rate as possible to maximize recovery.

**4.3.2 Overpressured Reservoirs.** One would intuitively expect overpressured gas reservoirs to be connected only to aquifers of a limited extent. However, the effects of aquifer-water influx on the performance history of overpressured reservoirs is well documented in the literature.[6,9,10] Dickinson[11] felt that most geopressured zones were found below the base of the main sands and were developed in or below a major shale series. The compaction and expulsion of water in the undercompacted shales adjacent to the producing interval could affect the performance history of an overpressured reservoir. Shale permeabilities are usually anisotropic because they are composed principally of flat-lying clay platelets. Vertical permeability is probably at least an order of magnitude less than the horizontal permeability.

A study by Bourgoyne *et al.*[12] indicated that the transmissibility of shales with a permeability of $10^{-5}$ md could be sufficient to affect the reservoir performance history. However, the shale-water effect would be insignificant if the shale permeability was reduced 100-fold to $10^{-7}$ md. The pressure profile across a sand/shale interface would be quite steep. Only the first few feet of the shale layer would be able to release the connate water before loss in permeability by compaction reduces the ability of the shale water to migrate in any significant volume. Measurements by Gondouin and Scala[13] showed shale permeability variations between $10^{-5}$ and $10^{-7}$ md.

Shale layers are areally extensive. However, water influx would be appreciable only in the areas of greatest pressure drawdown. A simulation study by Bass[9] amplified these findings. Shale-water influx occurs at early time and originates close to the wellbore because that is region of greatest pressure drop. Water influx is a function of the shale permeability as well as the degree and net thickness of the shale layers incorporated within or adjacent to the producing interval. Additionally, many overpressured gas sands display a high initial water saturation. Reservoir compaction could cause a portion of this water to become mobile. A portion of this mobilized water would migrate to producing wells.

Fetkovich *et al.*[1] included a relationship expressing the influence of water influx from an aquifer of limited extent and from interbedded shales on the effective-compressibility expression included in the overpressured gas reservoir material-balance equation. A limited aquifer is described as possessing $r_{eD} \leqq 5$ and $k \geqq 100$ md characteristics. The relative volume of interbedded shale is the ratio of the net productive thickness to the gross thickness. Both parameters are easily determined from well logs. History matching of apparent and calculated effective compressibility forms the basis for their solution technique. The variable-formation-compressibility problem is circumvented by assuming an average value.

## Nomenclature

$B_w$ = water formation volume factor, RB/STB
$c_e$ = effective compressibility, $Lt^2/m$, $psi^{-1}$

$c_f$ = formation compressibility, $Lt^2/m$, $psi^{-1}$
$c_g$ = gas compressibility, $Lt^2/m$, $psi^{-1}$
$c_w$ = water compressibility, $Lt^2/m$, $psi^{-1}$
$f(p,t)$ = pressure-/time-dependent potential function
$G$ = OGIP, $L^3$, scf
$G_p$ = cumulative gas produced, $L^3$, scf
$G_r$ = remaining gas volume, $L^3$, scf
$k$ = permeability, $L^2$, md
$p$ = pressure, $m/Lt^2$, psi
$p_a$ = pressure, $m/Lt^2$, psi
$p_i$ = initial pressure, $m/Lt^2$, psi
$\Delta p$ = pressure change, $m/Lt^2$, psi
$r_e$ = radius to outer reservoir limit, L, ft
$r_{eD}$ = $r_o/r_e$
$r_o$ = radius to water contact, L, ft
$r_D$ = dimensionless aquifer- to reservoir-radius ratio
$R_{s(aq)}$ = solution-gas/-oil ratio of aquifer water, $L^3/L^3$, scf/STB
$R_{s(cw)}$ = solution-gas/-oil ratio of connate water in shale, $L^3/L^3$, scf/STB
$S_{wi}$ = initial water saturation
$t$ = time, t,
$T$ = temperature, T, °R
$V$ = volume, $L^3$, $cm^3$
$V_p$ = PV, $L^3$, $cm^3$
$V_w$ = water volume, $L^3$, res bbl
$\Delta V_p$ = PV change
$W_e$ = volume of encroaching water, $L^3$, STB
$W_{e(sh)}$ = volume of encroaching shale water, $L^3$, STB
$W_{en}$ = net water influx, $L^3$, STB
$z$ = gas-deviation factor
$z_i$ = initial single-phase gas-deviation factor

## References

1. Fetkovich, M.J., Reese, D.E., and Whitson, C.H.: "Application of a General Material Balance for High-Pressure Gas Reservoirs," paper SPE 22921 presented at the 1991 Annual Technical Conference and Exhibition, Dallas, 6–9 October.
2. Hammerlindl, D.J.: "Predicting Gas Reserves in Abnormally Pressured Reservoirs," paper SPE 3479 presented at the 1971 SPE Annual Fall Meeting, New Orleans 3–6 October.
3. Stewart, P.R.: "Evaluation of Individual Gas Well Reserves," *Pet. Engineer* (May 1966) 85.

4. Harville, D.W. and Hawkins, M.F.: "Rock Compressibility and Failure as Reservoir Mechanisms in Geopressured Gas Reservoirs," *JPT* (July 1969) 1528.
5. Ramagost, B.P. and Farshad, F.F.: "*p/z* Abnormally Pressured Gas Reservoirs," paper SPE 10125 presented at the 1981 SPE Annual Technical Conference and Exhibition, San Antonio, Texas, 5–7 October.
6. Bernard, W.J.: "Gulf Coast Geopressured Gas Reservoirs," paper SPE 14362 presented at the 1985 SPE Annual Technical Conference and Exhibition, Las Vegas, Nevada, 22–25 September.
7. Prasad, R.K. and Rogers, L.A.: "Superpressured Gas Reservoirs: Case Studies and a Generalized Tank Model," paper SPE 16861 presented at the 1987 SPE Annual Technical Conference and Exhibition, Dallas, 27–30 September.
8. Bruns, J.R., Fetkovich, M.J., and Meitzen, V.C.: "The Effect of Water Influx on *p/z*-Cumulative Gas Production Curves," *JPT* (March 1965) 287.
9. Bass, D.M.: "Analysis of Abnormally Pressured Gas Reservoirs With Partial Water Influx," paper SPE 3850 presented at the 1972 SPE Abnormal Subsurface Pressure Symposium, Baton Rouge, Louisiana, 15–16 May.
10. Cronquist, C.: "Turtle Bayou 1936–83: Case History of a Major Gas Field in South Lousiana," *JPT* (November 1984) 1941.
11. Dickinson, G., "Geological Aspects of Abnormal Reservoir Pressures in Gulf Coast Lousiana," *AAPG Bull.* (February 1953) **37,** No. 2, 410.
12. Bourgoyne, A.T. *et al.*: "Shale Water as a Pressure Support Mechanism in Superpressure Reservoirs," paper SPE 3851 presented at the 1972 SPE Abnormal Subsurface Pressure Symposium, Baton Rouge, Louisiana, 15–16 May.
13. Gondouin, M. and Scala, C.: "Streaming Potential and the SP Log," *Trans.*, AIME (1958) **213**, 170.


## SI Metric Conversion Factors

| | | |
|---|---|---|
| ft | $\times 3.048^*$ | $E-01 = m$ |
| $ft^3$ | $\times 2.831\ 685$ | $E-02 = m^3$ |
| md | $\times 9.869\ 233$ | $E-04 = \mu m^2$ |
| psi | $\times 6.894\ 757$ | $E+00 = kPa$ |
| $psi^{-1}$ | $\times 1.450\ 377$ | $E-01 = kPa^{-1}$ |

*Conversion factor is exact.

# Chapter 5
# Use of Transient Pressures in the Material-Balance Equation

The material-balance equation is commonly used to evaluate gas reservoirs. Pressure relationships describing the interaction of the reservoir fluids are always related to static conditions. A well must be closed in for an extended time to ensure that pressure-buildup rate has stabilized. Shutting in a well not only reduces the cash-flow stream but, in many cases, also makes it difficult to bring a low-rate well back on production. Therefore, many operators hesitate to obtain static pressures because they feel the rewards are not worth the risks. A method that substitutes transient pressures for static pressures in the material-balance equation would be of considerable benefit to the industry because well downtime is reduced.

**Fig. 5.1** represents a typical bottomhole-pressure (BHP) buildup for a very-low-permeability well. The curve can be divided into three separate regions. Wellbore-inflow effects influence the shape of the curve and the pressure response in the early-time region. Equations describing pressure/rate relations in this region generally use the line-source, infinite-acting solution to the diffusivity equation. In the middle-time region, wellbore-inflow disturbances are negligible and the outermost reservoir boundary begins to affect the shape of the buildup curve. This portion of the buildup curve is a transition zone between early and boundary-dominated conditions. The late-time region occurs when the outer limit of the reservoir is the major influence on the pressure wave. The rate of pressure change is inversely proportional to the drainage pore volume (PV), and the pressure/time response is linear. The pseudosteady-state solution to the diffusivity equation describes the boundary-dominated fluid flow in this region.

The transient-*p/z* method was developed to substitute pressures obtained in the early part of the middle-time region for late-time static pressures. For instance, the new method permits use of transient pressures measured during the 100- to 500-hour period (Fig. 5.1) to estimate original gas in place (OGIP). Normal material-balance analysis methods would require use of pressure relationships obtained during the 1,500- to 3,000-hour span.

The concepts used in gaswell testing to determine the absolute-open-flow (AOF) potential provide the basis for this method. The calculated AOF potential is a measure of the productive capacity of a well. The gas-flow-rate equation developed by Rawlins and Schellhardt[1] from empirical data for evaluating low- to moderate-pressure wells is

$$q = C\left(p^2 - p_{wf}^2\right)^n, \quad (5.1)$$

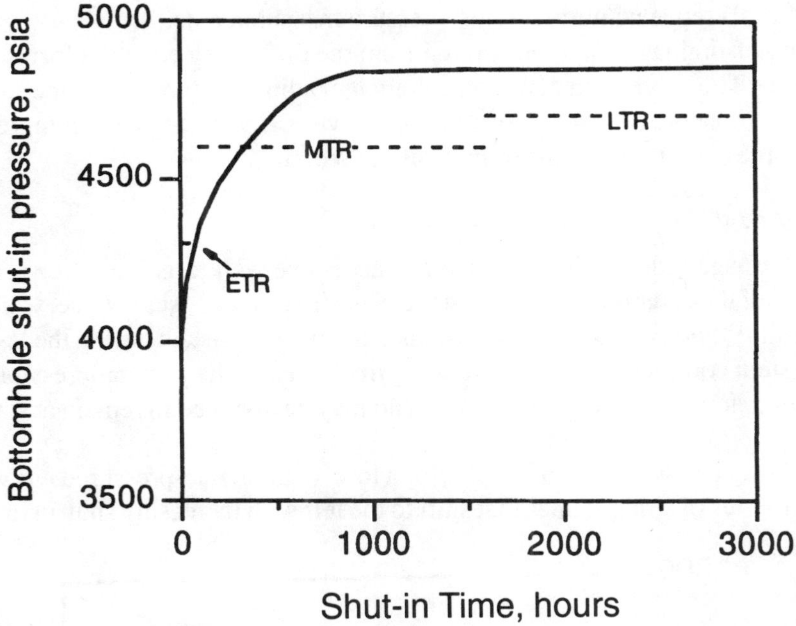


**Fig. 5.1—A typical pressure-buildup curve. The transient-*p/z* method shifts the pressure reference from the late-time region (LTR) to the early part of the middle-time region (MTR); ETR=early-time region.**

where $C$ = well-performance coefficient and the exponent $n$ defines the degree of laminar flow, with $n = 1$ for completely laminar-flow conditions. The well test is performed with a series of rate/flowing-pressure measurements. A logarithmic plot of test rate and the difference of the squares of the static reservoir and bottomhole flowing pressures (BHFP's) measured at the equivalent test rate is constructed from these data. The intercept and slope values of a straight line drawn through the data points are the well-performance coefficient and exponent terms. The AOF potential is calculated with Eq. 5.1 for the condition under which the BHFP is assumed to be equal to some base pressure. The exponent term is a function of the degree of ideal, or Darcy, flow, while the coefficient expression is a basket of terms defining the character of the well and reservoir. These terms include porosity, permeability, average viscosity/compressibility, reservoir thickness, and the radii of the outer and inner reservoir boundaries.

Cullender and Smith[2] originally developed the concept of replacing the static pressures in the AOF equation with transient, isochronal pressures. These authors theorized and showed in practice that the AOF could be calculated when short-time buildup pressures obtained from two or more well tests measured at some constant shut-in time (i.e., an isochronal interval) replaced normal static-pressure measurements.

The radius-of-drainage equation derived from the diffusivity equation forms the basis for their idea.[3] The expression defines the drainage radius as a function of the square root of time, reservoir porosity and permeability, gas viscosity, and system compressibility. System compressibility is discussed in detail in Sec. 6.1.2.

$$r_d = 0.029\sqrt{kt/\phi\mu c_s}\,. \qquad (5.2)$$

The same drainage radius is investigated by successive well tests if all the variables on the right side of the equation remain constant. Time is set at a constant value, while porosity, permeability, and gas viscosity are assumed to remain constant during the test period. Average system compressibility is governed principally by the gas-compressibility term in moderate- to low-pressure gas reservoirs and may be assumed to remain at some average value.

**Fig. 5.2** shows a typical isochronal plot for a low- to moderate-pressured gas well. The plot forms a series of straight lines that shift to the left with increasing shut-in time. Fluid

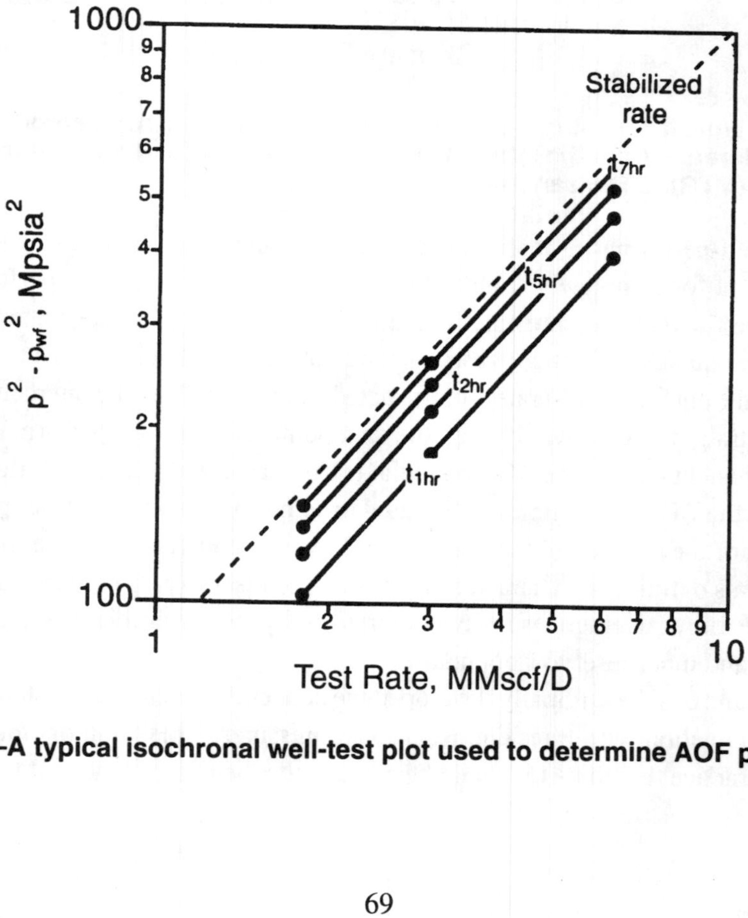


**Fig. 5.2—A typical isochronal well-test plot used to determine AOF potential.**

properties and formation compressibility are assumed to remain constant. These transient lines migrate toward the unknown static-pressure line. The BHFP's and rate relationships were measured at 1-, 2-, 5-, and 7-hour time intervals and at 1.8-, 3.0-, and 6.2-MMscf/D rates for this particular case. Note that the slopes of the lines are parallel and that the distance between the lines decreases with increasing shut-in time. The dashed line in the figure is the calculated relationship if static buildup pressures are measured during the test. Obviously, the time a well must be closed in is reduced if pressures obtained by the Cullender and Smith[2] isochronal testing method replace static pressures required for the Rawlins and Schellhardt[1] method.

## 5.1 Development of the Transient-*p/z* Concept

This method replaces material-balance static-pressure/fluid-property relationships with transient-pressure/fluid-property relations. Sullivan *et al.*[4] showed the transient-pressure *p/z* plot is able to calculate OGIP with a high degree of accuracy in certain cases. **Fig. 5.3** shows an example of the transient-*p/z* plot for the constant-rate case. The length of shut-in time increases in the following order, $t_s < t_1 < t_2$, where $t_s$ = time required for the well to attain static-pressure conditions. The distance between the isochronal-*p/z* line and the

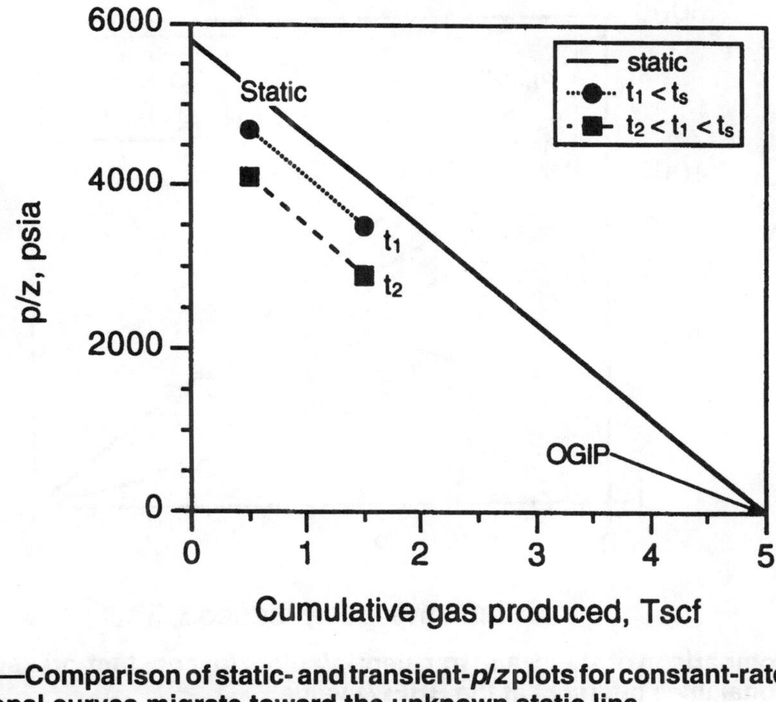


**Fig. 5.3—Comparison of static- and transient-*p/z* plots for constant-rate case. The isochronal curves migrate toward the unknown static line.**

static-*p/z* line decreases with increasing shut-in time. Sullivan *et al.* theorized that the transient-*p/z* plot has the same slope as a plot that uses stabilized pressure data as long as the shut-in times for each isochronal data point remain the same. The analysis procedure shifts the transient-*p/z* slope to intersect the known $(p/z)_i$ value on the *y* axis to emulate the unknown static-pressure *p/z* plot. The adjusted straight line is extrapolated to intersect the *x* axes to estimate the true OGIP. There would be no need to use static pressures in the *p/z* plot if the isochronal lines shown in Fig. 5.3 remain parallel over the producing history of the reservoir.

Subsequent work by Vu *et al.*[5] showed that the constant-pressure-depletion case behaves differently. **Fig. 5.4** indicates that the isochronal lines converge to the $(p/z)_{wf}$ value. The BHFP is 1,500 psia for this particular illustration. The procedure for estimating the OGIP is different from the procedure developed for the constant-rate case. A transient-*p/z* plot covering at least two points is constructed. The rightward extension of the line would end at the BHFP $(p/z)_{wf}$ value. A straight line is extrapolated from the static-$(p/z)_i$ value, through the $(p/z)_{wf}$ intersection point, and down to the *x* axis to estimate the true value of the OGIP. The quality of the interpretation procedure is enhanced by constructing two or more transient-*p/z* plots representing different shut-in times.

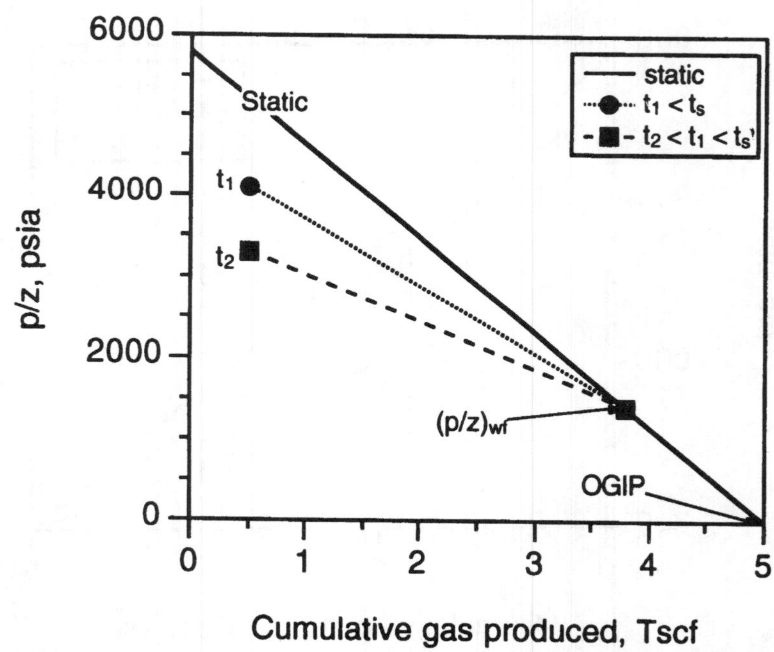


**Fig. 5.4—Comparison of static- and transient-*p/z* plots for constant-pressure case. The isochronal lines intersect at the BHFP/*z* value.**

Vu[6] provided mathematical justification for the transient-$p/z$ concept. The basic gaswell-testing equation (discussed by Dake[7] and Ramey and Cobb[8]) expressed in terms of pseudopressure forms the basic flow equation. The principle of superposition is included in the solution procedure for the case of a single-phase gas well located in the middle of a circular, homogeneous, isotropic reservoir. Skin effect is assumed to be insignificant. The following development assumes either constant-rate or constant-pressure-depletion conditions.

**5.1.1 Constant-Terminal-Rate Case.** The transient-line-source solution to the diffusivity equation for the constant-rate case expressed in terms of pseudopressures is

$$\frac{kh}{1,422qT}\left[m(p_i) - m(p_{wf})\right] = m_D(t_D) = \frac{1}{2}\left(\ln\frac{4t_D}{\gamma}\right). \quad (5.3)$$

The pseudosteady-state solution to the diffusivity equation under constant-rate boundary conditions is

$$\frac{kh}{1,422qT}\left[m(p) - m(p_{wf})\right] = \frac{1}{2}\left[\ln\frac{4A}{\gamma C_A (r_w)^2}\right]. \quad (5.4)$$

A pseudosteady-state solution to the material-balance equation is

$$\frac{kh}{1,422qT}\left[m(p_i) - m(p)\right] = 2\pi t_{DA}. \quad (5.5)$$

Dimensionless time, $t_{DA}$, is based on the drainage area; the $\mu c_g$-product term is evaluated at the average reservoir pressure, $p$; the system compressibility is assumed to be equivalent to the gas compressibility; and the rate term, $q$, is the final extended rate before shut-in.

Adding Eq. 5.4 to Eq. 5.5, using the principle of superposition to describe mathematically the pressure/rate/time relationships over the extended producing history of a well, and simplifying results in

$$\frac{kh}{1,422qT}\left[m(p_i) - m(p_{ws})\right] = m_D(t_D + \Delta t_D) - m_D(\Delta t_D). \quad (5.6)$$

Subtracting Eq. 5.4 from Eq. 5.6 expresses the relation in terms of static and shut-in pressures.

$$\frac{kh}{1,422qT}\left[m(p) - m(p_{ws})\right] = m_D(t_D + \Delta t_D) - m_D(\Delta t_D) - 2\pi t_{DA}. \quad (5.7)$$

The drawdown part of Eq. 5.7 is

$$m_D(t_D) = \frac{1}{2}\ln\frac{4A}{\gamma C_A (r_w)^2} - 2\pi t_{DA}, \quad \text{(5.8)}$$

and the buildup part is

$$m_D(\Delta t_D) = \frac{1}{2}\left(\ln\frac{4\Delta t_D}{\gamma}\right). \quad \text{(5.9)}$$

Eq. 5.7 can be expressed in a shortened form when the dimensionless production time is assumed to be much longer than the dimensionless shut-in time and the definitions shown as Eqs. 5.8 and 5.9 are included.

$$\frac{kh}{1,422qT}\left[m(p) - m(p_{ws})\right] = \frac{1}{2}\ln\frac{A}{C_A (r_w)^2 \Delta t_D}. \quad \text{(5.10)}$$

**5.1.2 Constant-BHFP Case.** The solution for the constant-terminal-pressure case expressed in terms of pseudopressures is[5]

$$\frac{\left[m(p_i) - m(p)\right]}{\left[m(p_i) - m(p_{wf})\right]} = \frac{56,538 G_p T}{Ah\phi(\mu c_g)\left[m(p_i) - m(p_{wf})\right]}. \quad \text{(5.11)}$$

Application of the principle of superposition accounts smoothly for the declining rate/time/pressure relationships. The $q_n$ term is defined as the last extended rate before shut-in.

$$\frac{kh}{1,422qT}\left[m(p_i) - m(p_{ws})\right] = \sum_{j=1}^{n}\left[\frac{(q_j - q_{j-1})}{q_n} m_D\left(t_{Dn} + \Delta t_D - t_{D(j-1)}\right)\right] - m_D(\Delta t_D). \quad \text{(5.12)}$$

Eq. 5.12 is altered to the following form when the definition for a constant dimensionless rate is included. Both sides of the equation are divided by $[m(p_i) - m(p_{wf})]$ and the producing history is assumed to be much longer than the shut-in time, resulting in

$$\frac{\left[m(p_i) - m(p_{ws})\right]}{\left[m(p_i) - m(p_{wf})\right]} = 1 - q_{Dn} m_D(\Delta t_D). \quad \text{(5.13)}$$

The superposition equation (Eq. 5.12) is combined with Eq. 5.13, shut-in time is assumed to be considerably less than producing time, and the buildup equation (Eq. 5.9) is substituted into Eq. 5.13, giving

$$\frac{kh}{1,422qT}\left[m(p) - m(p_{ws})\right] = \frac{1}{2}\ln\frac{A}{C_A(r_w)^2\Delta t_D}. \quad (5.14)$$

The form of Eq. 5.14 is the same as that of the constant-terminal-rate case (Eq. 5.10). The constant-rate and -pressure solutions to the diffusivity equation for transient-*p/z* conditions result in the same equation. Therefore, the concept is applicable for all field conditions.

## 5.2 Transient-*p/z* Plot

Eq. 5.10 or 5.14 can be rewritten in terms of *p/z* relationships when the Al-Hussainy *et al.*[9] approximation relating pseudopressures to gas properties and a finite-difference approximation are included in the equation.

$$\frac{kh}{1,422T}\left[\left(\frac{p}{z}\right) - \left(\frac{p}{z}\right)_{ws}\right] = \frac{\mu c_g q}{4}\ln\frac{A}{C_A(r_w)^2\Delta t_D}. \quad (5.15)$$

The difference between the static- and transient-*p/z* values on the left side of Eq. 5.15 is a function of the average $\mu c_g$ product for the constant-rate case, while the difference for the constant-pressure case would be $\mu c_g q$. The constant-pressure case also differs because the rate decreases while the $\mu c_g$ term increases. These inverse relationships cause a certain amount of compensation within the relative value of the total $\mu c_g q$ term. Additionally, the slope of the constant-pressure transient-*p/z* plot is not the same as the static-*p/z* plot because the rate term decreases to zero when the BHFP equals the reservoir pressure. For most instances, the logarithmic terms on the right side of the equation can be assumed to remain constant.

Eq. 5.15 may be rearranged to the following form when the $\Delta(p/z)$ term defines the difference between the *p/z* values at static and shut-in pressures and the dimensionless-real-time relationship are included.

$$\Delta\left(\frac{p}{z}\right) = \frac{355.5(\mu c_g)Tq}{kh}\ln\left[\frac{A\phi(\mu c_g)_{ws}}{0.000264C_A\Delta t}\right], \quad (5.16)$$

where $\mu c_g$ = average viscosity/compressibility product (Protos[10] found that the average gas viscosity and compressibility properties must be evaluated within the static- and transient-pressure ranges); $(\mu c_g)_{ws}$ = gas viscosity/compressibility term evaluated at $p_{ws}$; $q$ = gas flow rate; $\Delta t$ = difference between the time when the shut-in pressure was measured and the time required to attain the static pressure; and when $t$ is in hours and

$$t_D = \frac{A}{r_w^2} t_{DA} = \frac{0.000264kt}{\phi\mu c_g (r_w)^2} \quad (5.17)$$

The term on the left side of Eq. 5.16, $\Delta(p/z)$, remains constant if all the parameters on the right side of the equation remain constant. However, simulation studies showed this equality does not always hold true. The studies showed that the degree of nonparallelism between a static- and transient-*p/z* plot is a function of the difference between the gas properties evaluated at the shut-in and static pressures.

**5.2.1 Results of Simulation Studies.** ***Constant-Rate Case.*** Simulation studies by Sullivan *et al.*[4] and Vu *et al.*[5] showed that, in some instances, the transient-*p/z* plots become skewed for both the constant-rate and constant-pressure-depletion cases. The characteristics of the hypothetical reservoir simulated by Vu *et al.* were 160-acre radial drainage area, 20-ft producing-sand thickness, 5,000-psia initial pressure, 220°F reservoir temperature, and a 0.7 gas gravity. Volumetric OGIP was 3.594 Tscf. Depletion studies were conducted on 1.0-, 0.1-,and 0.01-md reservoirs. Wellbore storage and skin effects were not considered. The simulated reservoir was placed on production and periodically shut in to obtain both transient and static buildup pressures. The appropriate *z* factors were determined, and both static and transient-*p/z* plots were constructed from this information.

**Fig. 5.5** summarizes the results of the constant-rate studies. The transient- and static-*p/z* plots overlie each other even when the shut-in time was reduced to 7.5 minutes for the $k = 1.0$-md, $q = 90$-Mscf/D case study shown in Fig. 5.5a. Pressure drawdown is negligible, and there is little difference between the value of the $\mu c_g$-product term evaluated at either static or transient conditions. Extrapolation of any of the lines to 0 psia would result in the same OGIP estimate.

The results of the $k = 0.1$-md, $q = 90$-Mscf/D study shown in Fig. 5.5b indicate a small degree of skewness between the static- and transient-*p/z* plots. Shifting the 7.5-minute slope upward to intersect the $p_i/z_i$ intercept and extrapolating downward to the *x* axis results in a 13% underestimate of the volumetric OGIP estimate. The same shifting procedure conducted on the 30-minute transient-*p/z* plot reduces the error to 8%. The skewness is more pronounced for the $k = 0.01$-md case illustrated in Fig. 5.5c. In fact, the producing rate was decreased to 20 Mscf/D to provide a meaningful study. The results underestimate the OGIP by 24% if the 6-day transient-*p/z* plot replaces the static-*p/z* plot.

The $\mu c_g$-product term is the average value evaluated between the shut-in- and static-pressure limits. These studies show that skewness between the static- and transient-*p/z* plots is caused by the differences between the viscosity/compressibility-product term evaluated at transient and at static conditions.

***Constant-Pressure Case.*** **Fig. 5.6** summarizes the results of the three constant-pressure studies. The results indicate that the transient-*p/z* lines do not always converge to the point

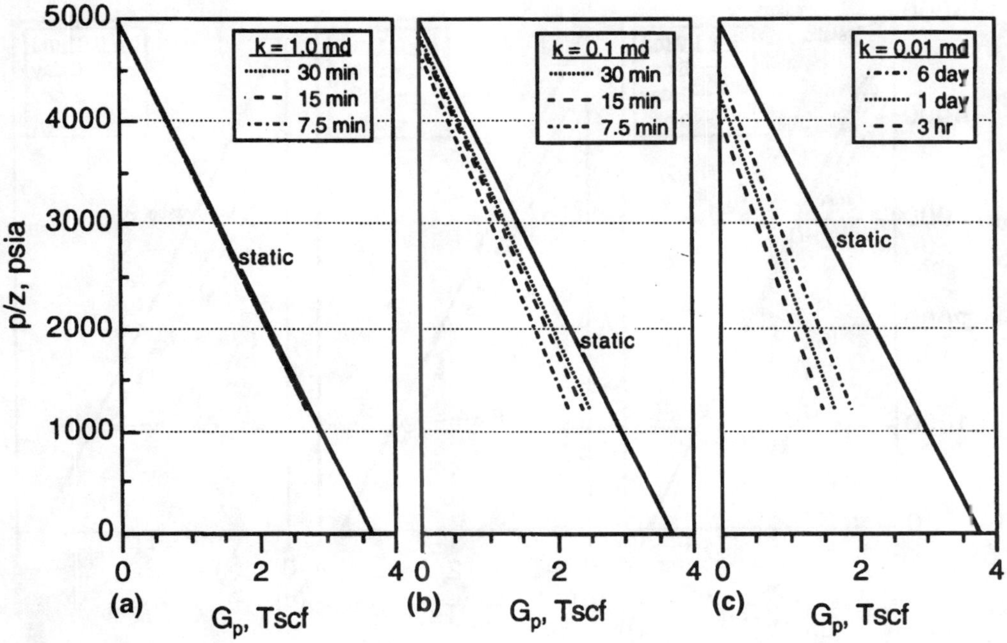


**Fig. 5.5—Simulation studies for constant-rate case; (a) $k$=1.0 md and $q$=90 Mscf, (b) $k$=0.1 md and $q$=90 Mscf/D, and (c) $k$=0.01 md and $q$=20 Mscf/D (from Ref. 6).**

where the $(p/z)_{wf}$ value is equal to the static-$p/z$ value. The $(p/z)_{wf}$ intersection is at 2,185 psia for the 1.0-md case study, producing at a constant $p_{wf}$ = 2,000 psia. Fig. 5.6a indicates a straight line constructed between $(p/z)_i$ and the 3-hour or 1- or 6-day $(p/z)_{wf}$ values are the same as in the static-$p/z$ plot. The concept of replacing static pressures with transient pressures works perfectly for this high-permeability case.

A constant $p_{wf}$ = 1,000-psia pressure was used in the two other studies. Extrapolation for the 0.1-md study shown in Fig. 5.6b of a line along the $(p/z)_i$ and either the 3-hour or 3- or 6-day $(p/z)_{wf}$ = 1,052-psia values results in an OGIP estimate that is 8% less than the static-$p/z$ estimate. This difference is well within the error incurred during normal data gathering.

The 3-hour transient-$p/z$ line does not coincide with the 3- or 6-day transient cases at the theoretical $p_{wf}/z$ = 1,052-psia point of intersection for the 0.01-md case study shown in Fig. 5.6c. The value of the transient-$\mu c_g$ term is not similar to the static-$\mu c_g$ value. Drawing a straight line from $(p/z)_i$ through the 3- and 6-day intersection results in an OGIP estimate that is only 12% less than the static-$p/z$ estimate. Note, that these numbers are somewhat less than those observed for the constant-rate case. The moderating differ-

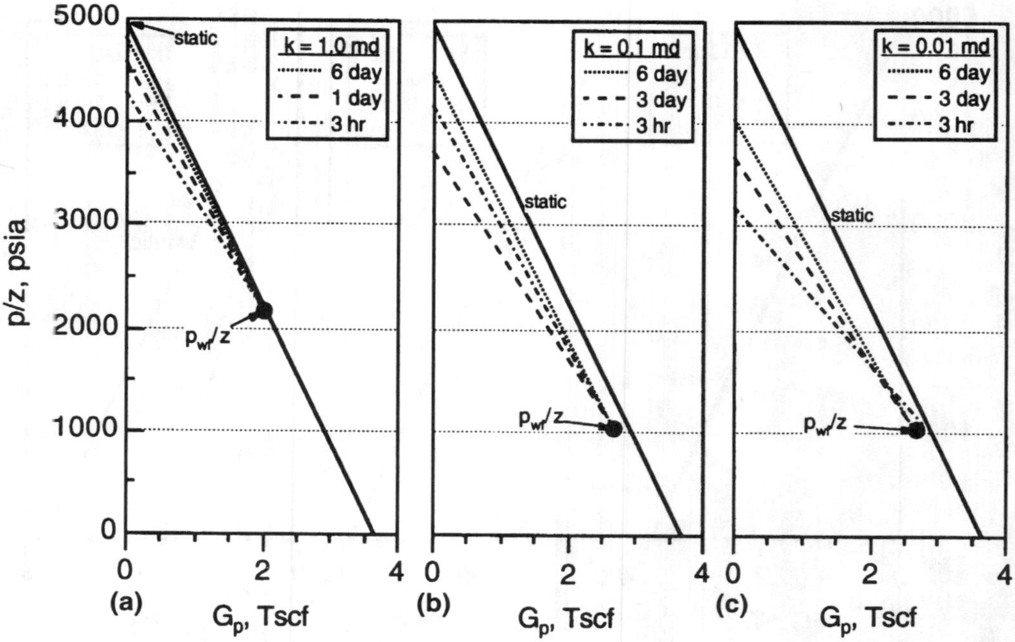


**Fig. 5.6—Simulation studies for constant-pressure case; (a) $k = 1.0$ md and $p_{wf} =$ 2,000 psia, (b) $k = 0.1$ md and $p_{wf} = 1{,}000$ psia, and (c) $k = 0.01$ md and $p_{wf} = 1{,}000$ psia (from Ref. 6).**

ence is caused by the compensating effect of multiplying the increasing viscosity/compressibility-product term with the decreasing producing rate.

Simulation studies show that the usefulness of the transient-*p/z* plot is a function of the degree of difference between $\mu c_g$-product term evaluated at transient and at static conditions. Therefore, the technique should work quite well for wells possessing a moderate drawdown but less well for low-permeability wells, which often produce at high drawdowns. However, the transient-*p/z* method will produce reasonable results if some preliminary effort is spent to study the drawdown effects before the buildup tests are analyzed.

## 5.3 Adjusting Shut-In Time

The previous discussion indicated skewness of the transient-*p/z* plots is caused principally by improper evaluation of the $\mu c_g$ term. **Fig. 5.7** is an example plot of the changes in the viscosity/compressibility-product term with pressure for a 0.7-gravity gas in a 220°F system. The change is linear at pressures $> 1{,}500$ psia. However, the slope of the $\mu c_g$-product term becomes increasingly nonlinear at pressures $< 800$ psia. Becerra-Ar-

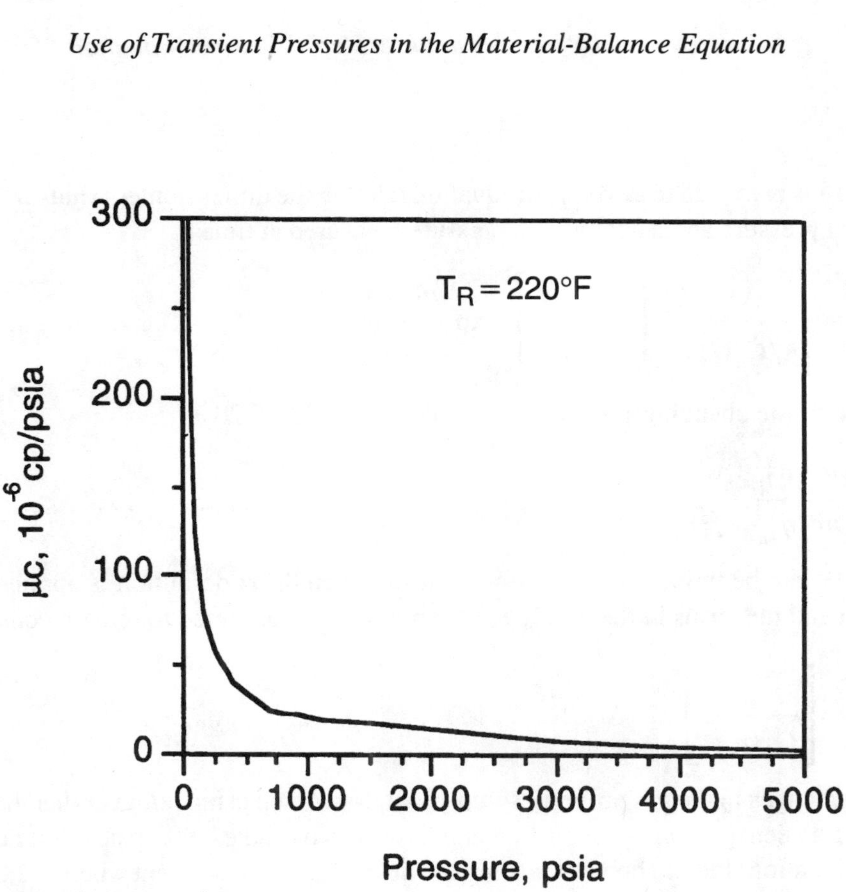


**Fig. 5.7—The value of the $\mu c_g$-product term changes dramatically at pressures below the 1,000 psia threshold.**

teaga[11] showed that these nonlinear effects are a function of gas gravity and temperature and are dramatically ameliorated for the case of overpressured gas reservoirs.

The work of Vu *et al.*[5] shows that the transient-*p/z* method can be applied when drawdown pressures remain close to the "linear" portion of the viscosity/compressibility curve and when the difference between the bottomhole static reservoir and flowing pressures is not significant. Transient-*p/z* plots become skewed from the static-*p/z* line when the pressure drawdowns become significant.

The left side of Eq. 5.15 defines the vertical distance between the static- and a particular transient-*p/z* line. Therefore, the following relationship must hold true for measurements at two or more points in the performance history.

$$\frac{(\mu c_g q)_1}{4}\ln\frac{A}{C_A(r_w)^2\Delta t_{D1}} = \frac{(\mu c_g q)_2}{4}\ln\frac{A}{C_A(r_w)^2\Delta t_{D2}} = \ldots\frac{(\mu c_g q)_n}{4}\ln\frac{A}{C_A(r_w)^2\Delta t_{Dn}}. \quad (5.18)$$

Eq. 5.18 is rewritten to develop an equation relating the dimensionless shut-in times of the initial pressure and a subsequent pressure measured at time *n*.

$$\Delta t_{Dn} = \frac{\Delta t_{D1}}{\left[A/C_A(r_w)^2\right]_i}\left[\frac{A}{C_A(r_w)^2}\right]_n \exp\frac{\left(\mu c_g q\right)_i}{\left(\mu c_g q\right)_n}. \qquad (5.19)$$

The ratio of the changing $\left(\mu c_g q\right)$ terms is defined in Eq. 5.20 as $a$.

$$a = \frac{\left(\mu c_g q\right)_i}{\left(\mu c_g q\right)_n}. \qquad (5.20)$$

Eq. 5.19 can be written in the following form when the $a$ definition is applied to the equation and the terms in the $A/C_A(r_w)^2$ expression are assumed to remain constant.

$$\Delta t_{Di} = \left[\frac{A}{C_A(r_w)^2}\right]^{1-a}\left(\Delta t_{D1}\right)^a. \qquad (5.21)$$

Small changes in the gas properties imply that drawdown is moderate or that the reservoir and transient pressures remain higher than an approximate 2,000-psia reference standard. Simulation studies showed that the skewness problem was absent when $>2{,}000$-psi pressure conditions were maintained. Trial calculations indicated $a \approx 1$ and $t_{D2} \approx t_{Di}$ for these conditions. Vu *et al.*[5] and Vu[6] used the same basic reservoir simulator described in the previous section to study the effect of applying the proposed adjustment procedure to account for the changing $\mu c_g$ term for both the constant-rate and constant-pressure cases.

**5.3.1 Results of Simulation Studies.** ***Constant-Rate Case.*** Applying Eq. 5.21 to account for changing shut-in time caused by the changing fluid properties was studied with the simulator and the following analysis procedure.

1. Use Eq. 5.17 to calculate dimensionless and real shut-in-time relations.
2. Apply Eq. 5.21 to calculate adjusted shut-in time for the expected pressure.
3. Run the simulator and measure pressures at adjusted shut-in times.
4. Plot the data to compare transient- and static-pressure results.

**Table 5.1** lists the basic data and results of the calculation procedure for the $k = 0.01$-md constant-rate simulation study. The pressures listed in Col. 1 should be viewed as estimates of the probable static BHP even though they correspond to calculated pressures for the simulated case. These estimates may be developed from field data by evaluating the bottomhole- or flowing-tubinghead-pressure/producing-rate history. Col. 2 lists the $\mu c$

**TABLE 5.1—REAL-/DIMENSIONLESS-TIME RELATIONSHIPS FOR CONSTANT-RATE* SIMULATION STUDY (from Ref. 6)**

| Pressure (psia) | $\mu c$ ($10^{-6}$ cp/psi$^{-1}$) | $a=(\mu c)_i/(\mu c)_n$ | $t_D$ | Shut-In Time (hours) |
|---|---|---|---|---|
| 4,709 | 3.75 | 1.000 | 338 | 3 |
| 4.436 | 4.02 | 0.933 | 629 | 6 |
| 4,180 | 4.29 | 0.874 | 1,083 | 11 |
| 3,933 | 4.59 | 0.817 | 1,838 | 20 |
| 3,702 | 4.90 | 0.765 | 2,965 | 35 |
| 3,481 | 5.22 | 0.718 | 4,577 | 57 |

*$q=20$ Mscf/D.
Other parameters: 3-hour initial shut-in time, $k=0.01$ md, $p_i=5{,}000$ psia, and $r_w=0.25$.

product evaluated at each pressure. A fluid-property curve similar to that in Fig. 5.7 was constructed. The following describes the calculation procedure shown in Table 5.1.

*Calculate Initial Dimensionless Time.* Use Eq. 5.17 to calculate the real-dimensionless-time relationship at initial conditions. An example calculation for the first timestep with a 3-hour initial shut-in time is

$$t_{Di} = \frac{0.000264kt}{\phi\mu c_g(r_w)^2} = \frac{(2.64\times10^{-4})(0.01)(3)}{(0.1)(3.75\times10^{-6})(0.25)^2} = 338.$$

*Calculate $\alpha$ Term for Projected Pressure Step.* Apply Eq. 5.20. The following example calculation is for the 4,436-psia-pressure step. The rate term is not included in the calculation because it remains constant for this case.

$$\alpha = \frac{(\overline{\mu c})_i}{(\overline{\mu c})_n} = \frac{3.75\times10^{-6}}{4.02\times10^{-6}} = 0.933.$$

*Adjust Initial Dimensionless Time To Account for Changing $\mu c$ Term.* Col. 2 of Table 5.1 lists the $\mu c$-product terms evaluated at the assumed static pressures listed in Col. 1. The shut-in time is adjusted with Eq. 5.21. The Dietz[12] shape factor of $C_A=31.6$ is for a well centered in a circular reservoir.

$$t_{D2} = \left[\frac{A}{C_A r_w^2 (r_w)^2}\right]^{1-a} \left(\Delta t_{Di}\right)^a = \left[\frac{(160\ \text{acres})(43,560\ \text{ft/acre})}{(31.6)(0.25)^2}\right]^{1-0.933} (338)^{0.933}$$

$$= 629.$$

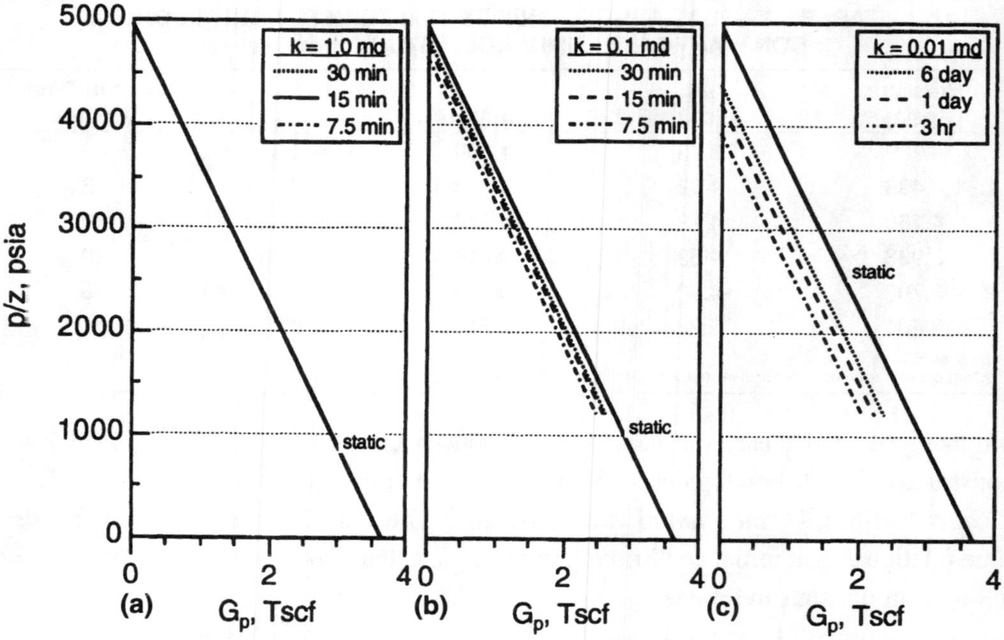


**Fig. 5.8—Constant-rate simulation studies with adjusted shut-in times; (a) $k$=1.0 md, (b) $k$=0.1 md, and (c) $k$=0.01 md (from Ref. 6).**

*Calculate Adjusted, Real Shut-In Time To Know When To Measure Pressure on New Buildup Curve.* The dimensionless-time equation (Eq. 5.17) is rewritten to convert the dimensionless time calculated in the previous step to real-time units.

$$t = \frac{t_D \phi \mu c_g (r_w)^2}{0.000264k} = \frac{(629)(0.1)(4.02 \times 10^{-6})(0.25)^2}{\left(2.64 \times 10^{-4}\right)(0.01)} = 6 \text{ hours}.$$

The increasing shut-in times required to compensate for the increasing $\mu c$ products are listed in the last column of Table 5.1 (see also **Fig. 5.8**). Note that the shut-in times increased from 3 hours (0.125 days) at the first measurement at 4,709 psia to 56.8 hours (2.4 days) after the reservoir pressure had declined to 3,481 psia. The ratio term is always $\alpha < 1$ and is constantly decreasing. Table 5.1 also shows that the value of the shut-in time increases at a faster rate than does the $\mu c$-product term. These relative changes imply that the radius of investigation does not remain constant but is changing with increased shut-in time, as would be expected. The quality of the calculated adjusted shut-in times was determined by entering these values into the reservoir simulator. Fig. 5.8c shows the results

**TABLE 5.2—REAL-/DIMENSIONLESS-TIME RELATIONSHIPS FOR CONSTANT-BHFP SIMULATION STUDY (from Ref. 6)**

| Pressure (psia) | Rate (MMscf/D) | $\mu c$ ($10^{-6}$ cp/psi$^{-1}$) | $a = (\mu cq)_i/(\mu cq)_n$ | $t_D$ | Shut-In Time (hours) |
|---|---|---|---|---|---|
| 4,562 | 28 | 3.89 | 1.000 | 324.5 | 3 |
| 4,227 | 25 | 4.24 | 1.028 | 251.2 | 2.5 |
| 3,945 | 22 | 4.56 | 1.086 | 146.3 | 1.6 |
| 3,530 | 20 | 5.13 | 1.062 | 183.1 | 2.2 |
| 3,241 | 18 | 5.60 | 1.081 | 153.5 | 2.0 |
| 2,932 | 16 | 6.02 | 1.131 | 96.2 | 1.4 |

*$p_{wf}$ = 1,000 psia.
Other parameters: 3-hour initial shut-in time, $k$ = 0.01 md, $p_i$ = 5,000 psia, and $r_w$ = 0.25.

for the 0.01-md, 3-hour study. The line is straight and parallel to the theoretical static-pressure condition.

Compare Fig. 5.8 with the unadjusted *p/z* lines shown in Fig. 5.5. All the transient-*p/z* lines for the $k$ = 1.0-, 0.1-, and 0.01-md, constant-rate cases are parallel to the static-*p/z* line. The simulation studies show that the time-adjustment procedure for the constant-rate case accounts for the potential skewness problem.

***Constant-Pressure Case.*** The reservoir simulator model was used to study the effect of adjusting shut-in time to compensate for the changing $\mu c_g q$ term for the constant-BHFP case. The calculation procedure is essentially the same as that for the procedure discussed for the constant-rate case except that it includes the producing rate in the numerator and denominator when calculating the $a$ term.

**Table 5.2** lists the reservoir data and results of the calculation procedure for the $k$ = 0.01-md, $p_{wf}$ = 1,000-psia case study. Col. 2 lists the average rate for that pressure step, while Col. 3 lists the $\mu c$ product evaluated at that pressure. The following describes the calculation procedure for the 3,250-psia-pressure step shown in Table 5.2.

*Calculate Initial Dimensionless Time.* Apply Eq. 5.17 to calculate the real-/dimensionless-time relationship at initial conditions.

$$t_{Di} = \frac{0.000264kt}{\phi\mu c_g(r_w)^2} = \frac{2.64 \times 10^{-4}(0.01)(3)}{(0.1)(3.89 \times 10^{-6})(0.25)^2} = 324.5 .$$

*Calculate $a$ Term for Projected Pressure Step.* Apply Eq. 5.20.

$$a = \frac{(\mu c q)_i}{(\mu c q)_n} = \frac{(3.89 \times 10^{-6})(28)}{(5.13 \times 10^{-6})(20)} = 1.062 .$$

*Adjust Initial Dimensionless Time To Account for Changing $\overline{\mu c_g}$ Term.* Eq. 5.21 adjusts the shut-in time. The Dietz[12] shape factor of $C_A = 31.6$ is for a well centered in a circular reservoir.

$$\Delta t_D = \left[\frac{A}{C_A(r_w)^2}\right]^{1-a}(\Delta t_{Di})^a = \left[\frac{(160\ \text{acres})(43,560\ \text{ft/acre})}{(31.6)(0.25)^2}\right]^{1-1.062}(324.5)^{1.062}$$

$$= 183.1.$$

*Calculate Adjusted, Real Shut-In Time To Know When To Measure Pressure on New Buildup Curve.* The dimensionless-time equation, Eq. 5.17, is rewritten to convert the dimensionless time calculated in the previous step to real-time units.

$$t = \frac{t_D \phi \mu c_g (r_w)^2}{0.000264k} = \frac{(183.1)(0.1)(5.13 \times 10^{-6})(0.25)^2}{2.64 \times 10^{-4}(0.01)} = 2.2\ \text{hours}.$$

Table 5.2 shows that the ratio term increases and $a > 1$. This observation implies that the changing rate term exerts a greater influence on the adjusted time than do the $\mu c$-product values. Recall that the $a$ term decreases for the constant-rate case.

**Fig. 5.9** illustrates the results of measuring the transient pressure at the calculated adjusted shut-in times for the three constant-pressure case histories shown in Fig. 5.6. For the most part, the skewness problem has been prevented. The problems that do arise are caused by the $\mu c$-product term being evaluated at $p < 2{,}000$ psia, where the fluid property changes are more dramatic than those that occur at higher pressures. However, in all cases, the skewness problem has been reduced significantly by this adjustment procedure.

## 5.4 Field Application

The general application of the transient-*p/z* technique to solve field cases may be hindered by the need for estimating a probable reservoir pressure to calculate a new shut-in time to compensate for the changing fluid properties for high-drawdown wells. Therefore, the field application of the adjusted-shut-in-time analysis procedure is divided into low- and high-drawdown categories. The following discussion relates the analysis criteria for low- to moderate-drawdown and high-drawdown reservoirs for constant-rate and constant-pressure-depletion cases.

The first step of any analysis is to construct a $\mu c_g$ vs. $p$ relationship to determine the shape and linearity of the curve over the expected pressure range. The plot will furnish a preliminary estimate of the expected significance of the gas-viscosity/-compressibility problem.

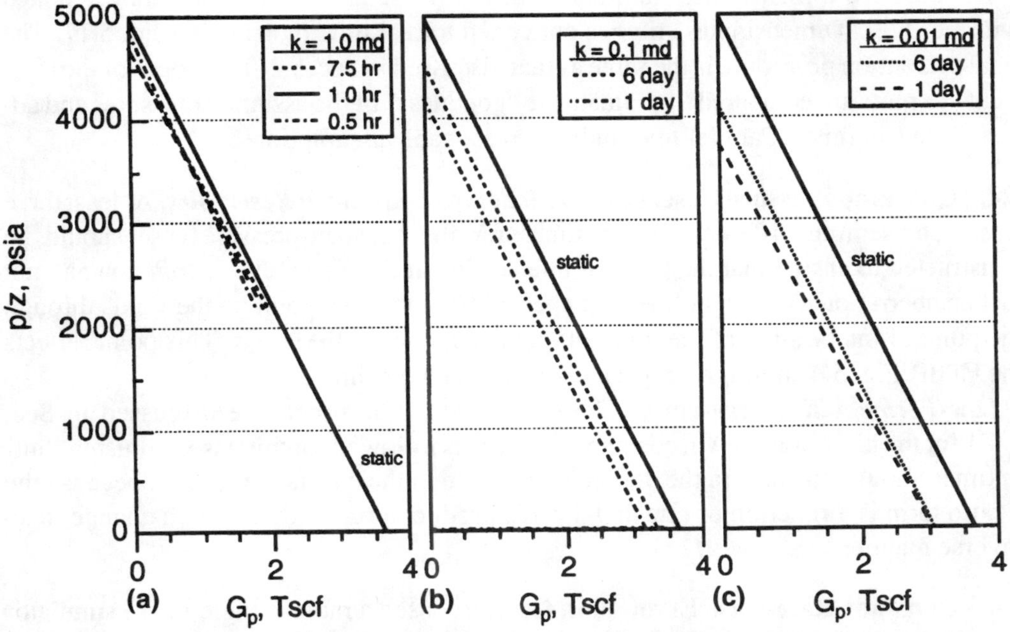


**Fig. 5.9—Constant-pressure simulation cases with adjusted shut-in times; (a) $k$=1.0 md, (b) $k$=0.1 md, and (c) $k$=0.01 md (from Ref. 6).**

**5.4.1 Constant-Rate Case.** ***Low- to Moderate-Drawdown Reservoirs.*** Transient-$p/z$ vs. $G_p$ values are calculated and plotted for two or more pressure buildups measured at two different periods of reservoir performance history. The degree of parallelism between two transient-$p/z$ lines representing different shut-in times provides a comparative method with which to establish the validity of the prediction method. The slope of the transient-$p/z$ line is shifted upward to intersect the $p_i/z_i$ value on the $y$ axis. This line represents the static-$p/z$ plot. Extrapolate downward to intersect the $x$ axis to estimate the OGIP. Refer to Fig. 5.3 to review the relationship between the transient- and static-$p/z$ plots for the constant-rate case.

***High-Drawdown Reservoirs.*** At least two transient-$p/z$ data points representing different shut-in times should be measured from the first buildup test curve. Calculate their respective dimensionless times with Eq. 5.17. A pressure must be assumed at the time of the next buildup test and the $\mu c_g$ values evaluated at the assumed pressure for each of the two initial values. The $\mu c_g$ and rate terms are entered into Eq. 5.20 to calculate the $a$ ratio values for the assumed pressure. Reservoir pressures may be estimated by noting changes in the flowing tubinghead pressure if the well is not producing liquids or if the liquid production remains essentially constant.

The adjusted-dimensionless and real shut-in-time relationship values are calculated with Eq. 5.21. Dimensionless time is converted to real time by applying Eq. 5.17. The final evaluation procedure is the same as that discussed in Sec. 5.2. The slopes of the transient-$p/z$ plots are compared to evaluate the "goodness" of the assumed pressures and adjusted shut-in times. Parallel lines indicate reasonable assumptions.

**5.4.2 Constant-Pressure Case.** ***Low- to Moderate-Drawdown Reservoirs.*** At least three lines representing different shut-in times for the constant-pressure case should be constructed to ensure that the point of intersection at the $p_{wf}/z$ value is common and not just an aberration. A straight line is connected from the $p_i/z_i$ point on the $y$ axis through the point of intersection of the transient-$p/z$ lines down to the $x$ axis. This point reflects the OGIP. Fig. 5.4 shows the expected shapes of the $p/z$ lines.

***Low-Permeability Reservoirs.*** The same calculation procedure discussed in Sec. 5.4.1 for the high-drawdown, constant-rate case is followed for this case. Adjusted shut-in times should not vary to the degree that is seen in the constant-rate case because the $a$ ratio term is a function of rate and the $\mu c_g$-product term, both of which change in an inverse manner.

**5.4.3 Constant-Rate-Case Error Plot.** Guerini[13] used a radial, single-phase simulator to model the behavior of a gas well located in the center of a reservoir containing either a 0.6- or 0.8-gravity gas. The model characteristics were similar to those of the Vu[6] model. Wellbore storage and formation damage were not considered in these constant-rate depletion studies. Various reservoir permeability, thickness, and producing-rate values were assigned to the simulator model to study the effect of pressure drawdown on quality of the transient-$p/z$ estimate of the OGIP.

The simulated well was periodically closed in during each depletion study to initiate a pressure buildup. Transient and static pressures were calculated from the buildup curve, and $p/z$ plots constructed. OGIP values obtained by the extrapolated static- and transient-$p/z$ analysis methods were compared, and the error of deviation between the two calculated. As expected, the deviation between static and transient OGIP values was found to be a function of drawdown.

Pressure drawdown is a function of transmissivity of the producing sand. Eq. 5.22 defines the error-plot correlation term. Viscosity and compressibility are assumed to remain constant.

$$E = kh/q, \qquad (5.22)$$

where $q$ is in Mscf/D.

An error vs. shut-in-time curve for various combinations of permeability, thickness, and gas flow rate was calculated. These values and curves are coalesced to a cohesive format

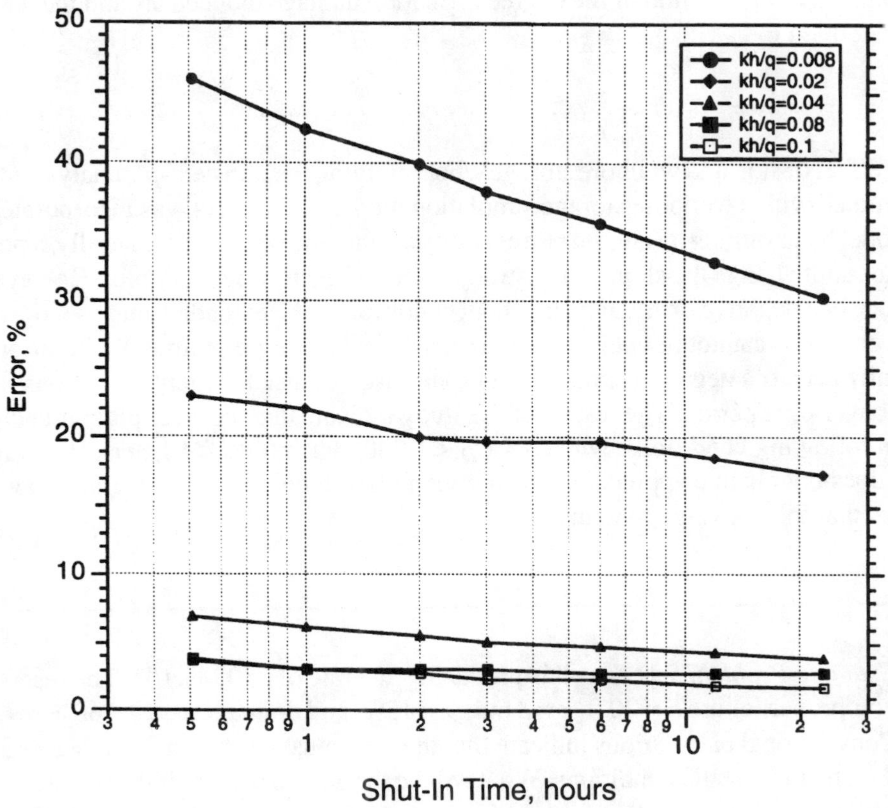


**Fig. 5.10—Error plot (from Ref. 3).**

by grouping the *kh/q* terms for each study onto the error plot (**Fig. 5.10**), remembering that rate units are Mscf/D. Each line in the figure represents the degree of error of the OGIP estimate for changing shut-in time for different transmissivity values. The error-chart values are not necessarily unique because the fluid properties are assumed to remain constant. There will be some variation according to character of the gas; but, for most cases, this variation will not alter the curves shown in Fig. 5.10.

Deshpande[14] included the effect of wellbore storage and skin in the derivation of the transient-*p/z* equation (Eq. 5.14). The dimensionless storage coefficient describes the significance of transient flow into the wellbore volume and is

$$C_D = \frac{0.894 V_{wb} T}{\phi h (r_w)^2 T_{wb}} . \quad (5.23)$$

The skin term is a function of the degree formation damage induced around the wellbore and is defined by

$$s = \frac{k_e - k_a}{k_a} \ln \frac{r_a}{r_w}. \quad (5.24)$$

The effect of skin and wellbore storage when applying the transient-$p/z$ analysis method was studied with a wellbore storage simulation model. Skin effect was incorporated into the model by assigning an altered permeability around the wellbore. Generally, a positive skin does not change the slope of the straight lines of the transient-$p/z$ plot. However, the presence of a negative skin causes transient conditions to be extended, and pseudosteady-state conditions cannot be achieved within a reasonable length of time. Wellbore storage generally acts as a negative-skin, altered-zone case because transient conditions remain for a longer time period. The results of the study indicate that the error plot will hold true for the following conditions: when $3 < C_D < 1{,}000$, when $0 < s < 10$, and when the $C_{De}$ value does not change appreciably during the pressure buildup. A moderate drawdown ensures that the $C_D$ value remains constant.

*Example Problem 5.1.* A well is producing at a rate of 750 Mscf/D. The reservoir properties are $h = 20$ ft, $k = 3$ md, $\phi = 15\%$, and drainage area $= 160$ acres. Conventional calculations indicate that the well must be shut in for at least 33 days to attain static conditions. What will be the degree of error if the transient-$p/z$ method is used with a 5-hour shut-in-time pressure to calculate OGIP?
*Solution.* The correlation parameter is

$$E = \frac{kh}{q} = \frac{(3)(20)}{750} = 0.08.$$

Reading the error chart in Fig. 5.10 indicates that the error would be approximately 3% if a 5 hour shut-in time replaced the 33-day pressure measurement.

The study shows that a pressure-depleting reservoir should act in a homogeneous manner and should have a permeability of $\geqq 0.1$ md for the error-chart technique to be applied. In reality, the technique appears to work in heterogeneous reservoirs if most of the producing section is composed of $k>0.1$-md rocks, if the well is not hydraulically fractured, and if wellbore storage effects are not long term. The method was applied with good results to a $k=6$-md reservoir when the well was fractured and a buildup test indicated $s=-3$.

*Example Problem 5.2.* Five pressure-buildup tests from the Mentor well were available for analysis.[15] Examination of the buildup curves indicated that pressures had not stabilized in the last four tests even after 70 hours of shut-in time. The average reservoir permeability calculated from pressure-buildup analysis was between 6 and 7 md. A hydraulic fracture treatment early in the well life changed the skin factor from +42 to −4, and a $kh=224$ md-ft was calculated.

*Solution.* The difference between the slope of the static- and transient-*p/z* lines in **Fig. 5.11** is minor. It is evident that the slope of the 30-minute transient-*p/z* line is essentially the same at the static-*p/z* line and will result in the same extrapolated OGIP answer. The 0.25- to 0.5-hour isochronal values for the prefracture time span are not parallel. This divergence is probably caused by a significant pressure drop induced by the initial skin problem. The large skin effect reduces the apparent formation permeability and causes the *kh/q* term to be reduced to less than the limiting value. Eq. 5.22 is used to calculate the correlation term for the Mentor well to evaluate the usefulness of the error-plot chart for determining the potential degree of error.

$$\frac{kh}{q}=\frac{224\ \text{md-ft}}{2.628\ \text{MMscf/D}}=0.085\ \text{md-ft/MMscf-D}.$$

Analysis of the error-plot chart indicates a <3% error for a 1-hour shut-in time. The actual error between the $p/z$ and $(p/z)_{1\text{hour}}$ extrapolations is calculated to be 1.8% when the simulated OGIP estimates are compared. Extrapolation along the static-pressure line to the *x* axis results in an estimate of OGIP = 19.8 Bscf. Shifting the slope of any of the isochronal lines to the $(p/z)_i$ intercept and extrapolating to the *x* axis would result in essentially the same ultimate recovery value, no matter which straight-line approximation is used. Application of the error chart analysis method to the transient-*p/z* analysis permits a 0.5-hour shut-in pressure to replace the 80-hour static-pressure tests when a <4% error is acceptable.

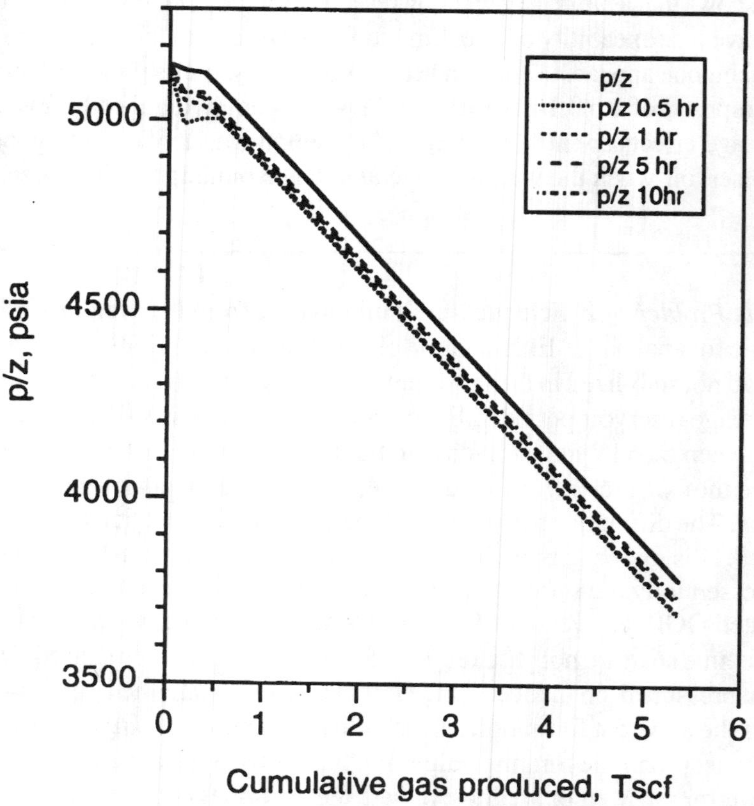


**Fig. 5.11—Comparison of static- and transient-*p/z* plots for the Mentor well (from Ref. 14).**

A practical approach for applying this method to field operations is to calculate a limiting production rate when a maximum error estimate is included with thickness and permeability estimates in a rearranged form of Eq. 5.22. The well production rate is reduced to the prescribed value for a few days to stabilize conditions before the buildup test is conducted. The radius of investigation equation (Eq. 5.2) approximates the time for a pressure disturbance to encounter boundary-dominated flow conditions.

**5.4.2 The Iterative-Slope Method.** All the terms in Eq. 5.16 remain constant except for the $\mu c_g$-product term. Simulation studies by Vu *et al.*[5] and Guerini[13] showed that the logarithmic nature of the $(\mu c)_{ws}$-product term allowed the value determined at any reasonable wellbore shut-in pressure to be a good approximation. However, evaluating the $\mu c_g$ expression presents a problem because the sought-after static reservoir pressure is one of the limits of the integrals for evaluating the average fluid-property value. Protos[10] developed an iterative scheme to calculate a most-likely value.

Error is introduced into calculation of the right side of Eq. 5.16 if there is a numerical error in an individual or a combination of values of the reservoir parameters. This percentage error in the $p/z$ estimates is included when an iterated calculation procedure is conducted over two or more shut-in times. For example, the $p/z$ values obtained by iteration should be in error by 20% if the total value of the errors in the constants is 20%. These errors are not completely compensating, but the total error in the final $p/z$ values for each iterated pressure should be less than the total error of the individual constants. Protos' simulation studies indicated that the final $p/z$ estimates from isochronal data tend to contain essentially the same percentage error.

If the previous discussion is true, then the line defined by two data points, although not true $p/z$ values, should exhibit very nearly the same slope as the unknown static-$p/z$ plot. Therefore, the slope of a line drawn between two iterated transient-$p/z$ data points may be shifted to intersect with the $(p/z)_i$ value on the abscissa to extrapolate to the OGIP.

***The Iterative Procedure.*** Eq. 5.25 defines the relationship between the shut-in point, $(p/z)_{ws}$, and the iterated value defined as $(p/z)^*$.

$$(p/z)^* = (p/z)_{ws} + \Delta(p/z). \qquad (5.25)$$

Protos' simulation study analyzed the calculation procedure for a worst-case scenario. **Fig. 5.12** illustrates the procedure applied to three 3-hour pressure buildups. The measured pressures are approximately 1,500 psi less than the static pressures. Data Set 1 covers an initial constant-rate case where the reservoir is depleted over the $p_1 \rightarrow p_2$ period. Data Set 2 covers the interval when the constant-pressure solution is used to deplete the reservoir during the $p_2 \rightarrow p_3$ period. Neither straight-line extrapolation is parallel to the true static-pressure $p/z$ line. The steps in the iterative/slope procedure are as follows.

1. Assume that the upper and lower limits of integration for calculating the average fluid properties are the initial and the shut-in pressures. Calculate an average of the two pressures to determine average values from the gas viscosity and compressibility terms. Multiply the two answers to determine the average $\mu c_g$ product. For practical purposes, a $\mu c_g$-product curve should work equally well.

2. Include the average viscosity/compressibility-product term obtained in Step 1 and the other estimated reservoir parameters in Eq. 5.16 to calculate a first estimate of $\Delta(p/z)$.

3. Add $\Delta(p/z)$ obtained in Step 2 to the value of the shut-in pressure to obtain a first estimate of $(p/z)^*$. Refer to Eq. 5.25.

4. Assume that the calculated $(p/z)^*$ term is the new upper limit of integration, and return to Step 1 to calculate new average viscosity and compressibility values. The $(p/z)^*$

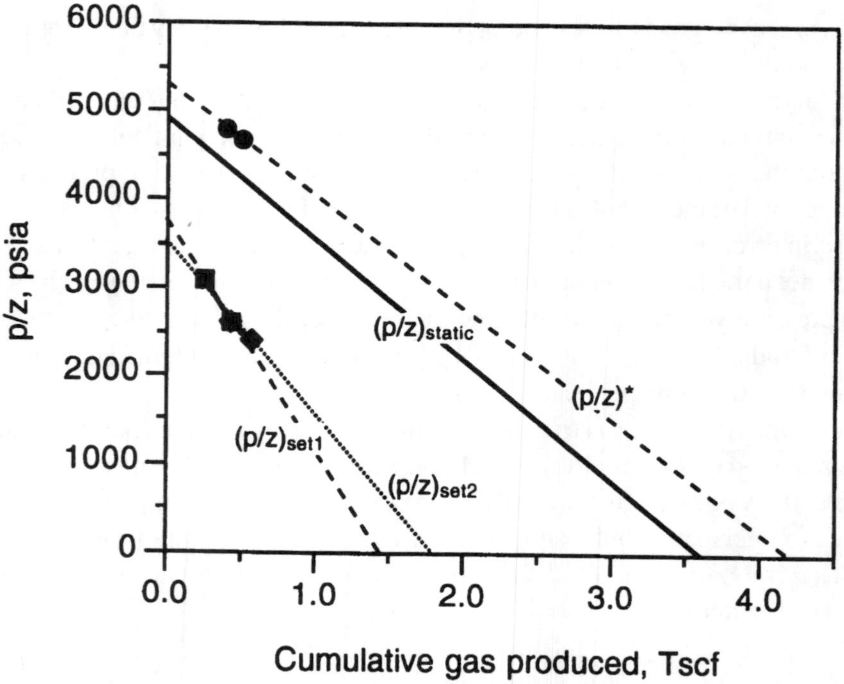


**Fig. 5.12—Illustration of how the iterative method permits use of three buildup pressures measured at different elapsed times and flow conditions to estimate OGIP (from Ref. 10).**

iterated value is correct if the original and calculated viscosity and compressibility values are similar. If not, go to Step 5.

5. Repeat Steps 1 through 5 until the difference between the input and outcome $(p/z)^*$ values are within a selected error limit.

Fig. 5.12 indicates the $(p/z)^*$ line is located above the static-$p/z$ line and the slope of the iterated-$(p/z)^*$ line is not exactly the same as the unknown, but theoretically correct, static-$p/z$ plot. However, the error in the OGIP estimate between the actual and iterated straight-line extrapolations is calculated as only 9.1% for this worst-case scenario.

***Ranges of Error.*** Protos[10] conducted studies to evaluate the degree of error incurred if the input values for the reservoir parameters are incorrect. Eq. 5.16 may be divided into a logarithmic and a $kh$-product sections. His simulation studies showed that all reservoir parameters lying within the logarithmic expression can vary by ±50% without materially affecting the results of the iterative process. This marginal effect is caused by the logarithmic compression of the values.

Calculation of the *kh*-product variables presents more of a problem. Simulation studies indicate that the error remains >5% but <9% if the *kh* term is overestimated by 25%. However, underestimating the *kh* term by 25% was found to change the results dramatically. Acceptable limits were attained for all cases after a 37-hour shut-in time. It is interesting to note that the percentage error flattens out and becomes stabilized with time. This effect is caused by the shut-in pressure approaching the unknown static-pressure value.

Application of the iterative method to determine remaining reserves in a pressure-depleting gas reservoir may be summarized as follows.

1. The reservoir permeability and thickness estimate should be within 25% of the actual value.
2. The estimate of the drainage area should be within 50% of the actual value.
3. The reservoir permeability should be >0.1 md.
4. The technique is applicable to extremely-low-permeability ($k=0.01$ md) wells if the reservoir permeability, thickness, and drainage area are known with reasonable accuracy.
5. The technique is not applicable for low- to intermediate-conductivity, hydraulically fractured wells producing from low-permeability reservoirs.

*Example Problem 5.3—Quaker State Well.* A single 82-hour pressure-buildup test was available for analysis. Ref. 16 gives the following relevant data that were obtained from pressure records, log and core data, and general knowledge: $G_p=1.171$ Bscf at the time of shut-in, $A=640$ acres, $h=10$ ft, $\phi=8.75\%$, $p_i=3{,}001$ psi, and $p_{ws}=1{,}866$ psi. Extrapolation of the middle-time region of the pressure-buildup test resulted in a permeability estimate of 14 md. The average reservoir pressure was calculated from the extrapolation of the well-testing $p^*$ term to be 2,293 psia. A measured buildup pressure of 1,866 psia at a shut-in time of 12 hours was read from the buildup test. The iterated $(p/z)^*$ was calculated to be 2,271 psi, which is essentially coincident with the static pressure measured from the extended buildup test. OGIP estimates between static- and iterative-derived $p/z$ values were all within a 1% error.

*Solution.* Measured pressure-buildup values at 6 and 12 hours are used to calculate a value for the apparent average static reservoir pressure. The iteration process is conducted until the pressure difference between the two buildup times is within a 1% limit. The iterative procedure applied to 6- and 12-hour pressure values results in essentially the same OGIP value as that calculated from the long-term buildup. For all practical purposes, the OGIP values produced from the 1-, 6-, and 12-hour pressure measurements produce essentially the same static-reservoir-pressure value as is determined from the extended buildup calculation.

## Nomenclature

$A$ = area, $L^2$, $ft^2$
$B_w$ = water formation volume factor, $L^3/L^3$, RB/STB
$c_e$ = effective compressibility, $Lt^2/m$, $psi^{-1}$
$c_f$ = formation compressibility, $Lt^2/m$, $psi^{-1}$
$c_g$ = gas compressibility, $Lt^2/m$, $psi^{-1}$
$c_s$ = system compressibility, $Lt^2/m$, $psi^{-1}$
$c_w$ = water compressibility, $Lt^2/m$, $psi^{-1}$
$C$ = well performance coefficient
$C_A$ = Dietz shape factor
$E$ = error-plot correlation term
$f(p,t)$ = pressure-/time-dependent potential function
$G$ = OGIP, $L^3$, scf
$G_p$ = cumulative gas produced, $L^3$, scf
$G_r$ = remaining gas volume, $L^3$, scf
$m$ = mass, m
$m(p)$ = general pseudopressure term
$k$ = permeability, $L^2$, md
$p$ = pressure, $m/Lt^2$, psi
$p_a$ = abandonment pressure, $m/Lt^2$, psi
$p_i$ = initial pressure, $m/Lt^2$, psi
$p_{wf}$ = bottomhole flowing pressure, $m/Lt^2$, psi
$p^*$ = iterated pressure, $m/Lt^2$, psi
$\Delta p$ = pressure change, $m/Lt^2$, psi
$q$ = gas flow rate, $L^3/t$, scf/D
$r_d$ = drainage radius, L, ft
$r_e$ = radius to reservoir limit, L, ft
$r_o$ = radius to water contact, L, ft
$r_w$ = wellbore radius, L, ft
$R_D$ = dimensionless aquifer- to reservoir-radius ratio
$R_{s(aq)}$ = solution-gas/-oil ratio of aquifer water, $L^3/L^3$, scf/STB
$R_{s(cw)}$ = solution-gas/-oil ratio of connate water in shale, $L^3/L^3$, scf/STB
$S_{wi}$ = initial water saturation
$t$ = time, t,
$t_s$ = time required to attain static-pressure conditions, t,
$T$ = temperature, T, °R
$T_R$ = reservoir temperature, T, °F
$V$ = volume, $L^3$, $cm^3$

$V_p$ = PV, $L^3$, $cm^3$
$V_w$ = water volume, $L^3$, res bbl
$\Delta V_p$ = PV change
$W_e$ = volume of encroaching water, $L^3$, STB
$W_{e(sh)}$ = volume of encroaching shale water, $L^3$, STB
$W_{en}$ = net water influx, $L^3$, STB
$z$ = gas-deviation factor
$z_i$ = initial single-phase gas-deviation factor
$\alpha$ = ratio of changing $q(\mu c_g)$ terms
$\gamma$ = Euler's constant = 0.5772
$\mu$ = viscosity, m/Lt, cp
$\phi$ = porosity

**Subscripts**

$D$ = dimensionless
$i$ = at initial conditions
$n$ = at later-time conditions
$wf$ = bottomhole flowing
$ws$ = shot in
$p$ = at any pressure

## References


1. Rawlins, E.L. and Schellhardt, M.A.: *Backpressure Data on Natural Gas Wells and Their Application to Production Practices*, Monograph, U.S. Bureau of Mines, Washington, DC (1936) **7.**
2. Cullender, M.H. and Smith, R.H.: "The Isochronal Performance Method of Determining the Flow Characteristics of Gas Wells," *Trans.*, AIME (1955) **202,** 137.
3. Earlougher, R.C. Jr.: *Advances in Well Test Analysis*, Monograph Series, SPE, Richardson, Texas (1977) **5,** 264.
4. Sullivan, S.A., Poston, S.W., and Piper, L.A.: "Using Short-Term Pressure-Buildup Tests for Reserves Estimation in Tight-Gas Reservoirs," paper SPE 17707 presented at the 1988 SPE Gas Technology Symposium, Dallas, 13–15 June.
5. Vu, T., Chen, H.Y. and Poston, S.W.: "Using Short-Term Pressure Tests To Calculate Gas Reserves," paper SPE 20111 presented at 1990 SPE Permian Basin Oil and Gas Recovery Conference, Midland, Texas, 8–9 March.
6. Vu, T.: "Using Short Term Pressure Buildup Tests to Calculate Gas Reserves," MS thesis, Texas A&M U., College Station, Texas (May 1990).
7. Dake, L.P.: *Fundamentals of Reservoir Engineering*, Elsevier Scientific Publishing Co., New York City (1978).

8. Ramey, H.J. Jr. and Cobb, W.M.: "A General Pressure-Buildup Theory for a Well in a Closed Drainage Area," *JPT* (December 1971) 1493; *Trans.*, AIME, **251.**
9. Al-Hussainy, R., Ramey, H.J. Jr., and Crawford, P.B.: "The Flow of Real Gases Through Porous Media," *JPT* (May 1966) 624; *Trans.*, AIME, **237.**
10. Protos, N.E.: "Application of the Isochronal, Transient *p/z* Plotting Method for Determination of Original Gas In Place to Low Permeability Reservoirs," MS thesis, Texas A&M U., College Station, Texas (August 1991).
11. Becerra-Arteaga, T.: "Analysis of Abnormally Pressured Gas Reservoirs," PhD dissertation, Texas A&M U., College Station, Texas (May 1993).
12. Dietz, D.N.: "Determination of Average Reservoir Pressures From Buildup Surveys," *Trans.*, AIME (1965) 955.
13. Guerini, A.F.: "Criteria to Determine Gas Reserves Using Short Term Build-Up Tests," MS thesis, Texas A&M U., College Station, TX (December 1991).
14. Deshpande, A.: "Applying the Isochronal, Transient *p/z* Method to Gas Wells Exhibiting Wellbore Storage Effects," MS thesis, Texas A&M U., College Station, Texas (May 1993).
15. Poston, S.W. and Chen, H.Y.: "Using Short Term Pressure Transients to Determine Reserves and Reservoir Character," milestone report, Contract 91-0273, Gas Research Inst., Chicago, Illinois (April 1992).
16. Poston, S.W., Chen, H.Y. and Deshpande, A.: "Short-Term Test Can Cut Cost of Gas Reserve Estimation," *Oil & Gas J.* (14 September 1992) 85.

## SI Metric Conversion Factors

| | | | |
|---|---|---|---|
| acre | × 4.046 873 | E − 01 | = ha |
| bbl | × 1.589 873 | E − 01 | = $m^3$ |
| cp | × 1.0* | E − 03 | = Pa·s |
| ft | × 3.048* | E − 01 | = m |
| $ft^3$ | × 2.831 685 | E − 02 | = $m^3$ |
| °F | (°F − 32)/1.8 | | = °C |
| md | × 9.869 233 | E − 04 | = $\mu m^2$ |
| psi | × 6.894 757 | E + 00 | = kPa |
| $psi^{-1}$ | × 1.450 377 | E − 01 | = $kPa^{-1}$ |

*Conversion factor is exact.

# Chapter 6
# Estimating Original Gas in Place and Predicting Performance

## 6.1 Introduction

Direct and iterative-type models have been developed to study overpressured gas reservoirs. The direct-input model requires some prior knowledge of formation-rock and -fluid properties. The calculations are usually compared with the properties of a straight line to determine the quality of the answer. Iterative models generally determine a single variable that describes the depletion style of the reservoir. The value of this all-inclusive term is calculated by assuming various original-gas-in-place (OGIP) values and determining which value is the best fit to a line corresponding to a predetermined set of conditions.

**6.1.1 Direct-Input Models.** Hammerlindl[1] developed two methods for comparing the compressibility effects for an overpressured, depletion-drive reservoir with those for a normal pressure situation. The coefficient calculated by either method provides a factor for relating the apparent OGIP determined from the incorrect, early-time extrapolation of the $p/z$ curve to the true value. Ramagost and Farshad[2] rewrote the material-balance equation to include the effect of formation and connate-water compressibility for an overpressured, depletion-drive reservoir. A straight line results when the correct gas, water, and formation-compressibility terms are included in the material-balance equation describing the performance history. Formation and water compressibility were assumed to remain constant.

Wang and Teasdale[3] coupled a material-balance model for an overpressured gas reservoir with a water-influx model to study pressure-depleting and waterdrive reservoirs. Although the concept is useful, the program is not available, it must be installed in a personal computer, and the operator must have knowledge of its operation. Begland and Whitehead[4] used a pressure-depletion, material-balance model to study the pressure-dependent effects of formation compressibility and water solution gas on the calculated OGIP. They concluded that OGIP could be overestimated by 25% if these effects were not taken into account.

Roach[5] initially developed the most practical method for studying overpressured, pressure-depleting gas reservoirs. The material-balance equation was rearranged in the form of an equation of a straight line. The input variables are production/pressure/gas-deviation-factor relationships, while the outcome variables are OGIP and a single-valued expression for the formation compressibility. The utility of the method was not appreciated until Poston *et al.*[6] showed that the curvature of the data plot reflects the type, strength, and time frame over which transient, steady-state, and water-influx conditions occurred.

**6.1.2 Iterative or Curve-Fitting Models.** Bernard[7] presented a model based on a least-squares approximation for estimating a coefficient accounting for the effects of rock and water compressibility. Analysis of 13 overpressured gas reservoirs in Louisiana indicated that the true OGIP could be overestimated by 70% when a straight-line extrapolation of the early-time data was used. He felt that all these so-called pressure-depleting reservoirs experienced either shale-water or minor aquifer-water influx. Ambashtha[8] applied a type-curve matching approach to study overpressured, depletion-drive gas reservoirs. The results are somewhat problematical because only one reservoir example is discussed.

Fetkovich *et al.*[9] modified the material-balance equation to calculate the variable-effective-formation-compressibility vs. pressure curve for various assumed values of OGIP. An alternative apparent-formation-compressibility vs. pressure term is also calculated for various aquifer sizes. Similarity of the two curves indicates an acceptable answer for OGIP and aquifer size.

The vast majority of the industry does not have the time nor the expertise to use the specialized iterative techniques to determine OGIP and predict future performance. Therefore, the following discusses the characteristics of the direct-input models that can be easily arranged for solution with a spreadsheet program.

## 6.2 Hammerlindl Single-Point Adjustment Method

The *p/z* plot of an overpressured gas reservoir usually exhibits different slopes at early and late times. The initial, lesser-valued slope reflects the sum of gas-, rock-, and water-compressibility values evaluated at high pressures, while the later, steeper slope observed at moderate to low reservoir pressures reflects the singular effect of the greatly increased value of the gas-compressibility term. Hammerlindl[1] recognized the curved nature of the overpressured *p/z* plot and developed two methods for adjusting the early-time slope to a more realistic value. Ambastha[10] derived the fundamental relationships shown in the Hammerlindl paper.

**6.2.1 Concept Development.** Hammerlindl[1] adjusted the early-time *p/z* data to reflect the effects of connate water and formation compressibility. Variable-formation-compressibility and water-influx drive mechanisms are assumed to be absent. The main attribute of either of the two Hammerlindl methods is the ability to adjust very early-time production/pressure data to the expected normal-pressured, steeper-slope conditions. However, the results of either adjustment procedures should be viewed with caution because of inherent inaccuracies. These inaccuracies are (1) whether the proper expression for the formation compressibility is known, (2) whether water influx is present and in what strength, (3) the proper normal-pressured value to use as the endpoint, and (4) the quality of the early-time information. These adjustment procedures should be used only when nothing else is available, as is often the case. The following briefly discusses the development of each method.

**6.2.2 Average-System-Compressibility Method.** The ratios of the system compressibility to gas compressibility evaluated at initial and assumed-normal-pressured conditions are averaged to calculate a correction factor. This factor is then used to correct the OGIP overestimate of the early-time *p/z* plot.

$$F_c = \frac{(c_s/c_g)_i + (c_s/c_g)_n}{2} = \left(\overline{c_s/c_g}\right). \quad (6.1)$$

The system-compressibility equation is derived from material-balance concepts discussed by Craft and Hawkins.[11] The term reflects the compressibility of the mobile phase and is the sum of the compressibility effects of the formation, connate-water, and gas phases of the system. Their definition for the system compressibility is

$$c_s = \frac{c_{gi}S_{gi} + c_w S_{wi} + c_f}{S_{gi}}. \quad (6.2)$$

The OGIP estimate obtained from extrapolating the early-time *p/z* data, $G_a$, is divided by the correction factor to account for the influence of the larger gas-compressibility value at normal reservoir pressures. The correction factor is always $F_c \geqq 1$.

$$G = \frac{G_a}{F_c}. \quad (6.3)$$

**6.2.3 Change-in-Reservoir-Volume Method**. Hammerlindl[1] applied material-balance concepts to develop a ratio between the slopes of normal-pressured and overpressured *p/z* straight-line extrapolations. The correction factor is the ratio between the difference in the gas formation volume factor at normal-pressured conditions to that at overpressured conditions. Gas compressibility, defined in terms of the gas formation volume factor and effective formation compressibility, are considered for this case. The subscripts define values measured at initial and normal-pressured conditions. The adjustment term is always $G_{pr} \leqq 1$ because the effective-formation-compressibility term is included in the denominator.

$$G_{pr} = \frac{B_{gi} - B_{gn}}{B_{gi} - B_{gn} + \left\{\left[B_{gn}(p_i - p_n)\left(c_f + S_{wi}c_w\right)\right]/(1 - S_{wi})\right\}}. \quad (6.4)$$

The correction factor is multiplied by the value obtained by the OGIP extrapolation of the early-time *p/z* data to provide the "true" OGIP value.

Either Hammerlindl method requires knowledge of the water- and formation-compressibility values. Ref. 1 presents a correlation relating formation compressibility to depth; Fig. 2.11 shows the correlation. This correlation should be avoided if at all possible because of the paucity of data points.

*Example Problem 6.1.* Hammerlindl[1] illustrated the two previous solution methods with a study of the NS2B Unit in the North Ossum field in Lafayette Parish, Louisiana. Harville[12] and Harville and Hawkins[13] published the data in an earlier work. The initial reservoir pressure of the limy, fine-grained sand was 8,921 psia at 12,500 ft. The 0.714-psi/ft gradient is in excess of the 0.465-psi/ft normal pressure gradient. **Table 6.1** lists the reservoir data. Fluid properties of the gas-condensate reservoir were developed from laboratory fluid-analysis reports.

Faults and the presence of a downdip water contact define the structure of the unit. Structural control indicated an aquifer of very limited extent. Water production was negligible during depletion of approximately 35% of the expected reserves. Volumetric studies calculated an OGIP of 114 Bscf. **Fig. 6.1** presents the $p/z$ plot reflecting the production history. Extrapolation of the early-time $p/z$ data to 0 psia results in an OGIP estimate of 220 Bscf, while the late-time extrapolation indicates an OGIP of 130 Bscf. The evaluation problem was to reconcile the differences between the volumetric and $p/z$ plot estimates.

A second and normal gradient reservoir pressure must be selected to apply the method. Hammerlindl selected a 6,951-psia measured pressure translating to a 0.55 psi/ft gradient. Craft and Hawkins[11] state that hydraulic gradients vary from 0.50 psi/ft for brine to 0.433 psi/ft for freshwater reservoirs. The earlier reservoir study by Harville and Hawkins[13] indicated that formation compressibility had decreased to a more normal value when production caused the reservoir pressure gradient to decline to 0.55 psi/ft. Normal gradients along the U.S. gulf coast are generally assumed to be 0.465 psi/ft.

*Solution.* The first part of the solution is to use Fig. 2.11 or some other correlation to determine the formation-compressibility value at the 12,500-ft reservoir datum. The figure indicates that, at 12,500 ft, $c_f = 21.9 \times 10^{-6}$ psi$^{-1}$. Apply the average-effective-system-compressibility method. Adjust the results of the analysis of the early-time $p/z$ curve to account for the effect of rock and water compressibility on the production/pressure response. Reservoir data are included in Eq. 6.2 to determine the initial, overpressured system compressibility at 8,921 psia.

$$c_s = \frac{(30 \times 10^{-6})(0.66) + (3 \times 10^{-6})(0.34) + (21.9 \times 10^{-6})}{0.66}$$

$$= 64.7 \times 10^{-6} \text{ psi}^{-1}.$$

| TABLE 6.1—N2SB RESERVOIR DATA (from Ref.1) | |
|---|---|
| $p_i$, psia | 8,921 |
| $p_2$, psia | 6,921 |
| $p_{DP}$, psia | 6,900 |
| $B_{gi}$, scf/ft$^3$ | 302.5 |
| $c_{gi}$, psi$^{-1}$ | $30\times10^{-6}$ |
| $c_{g2}$, psi$^{-1}$ | $48.7\times10^{-6}$ |
| $c_w$, psi$^{-1}$ | $3\times10^{-6}$ |
| $B_{g2}$, scf/ft$^3$ | 281.5 |
| $\phi$, % | 23 |
| $S_{wi}$, % | 34 |
| $T_R$, °F | 250 |
| $z_i$ | 1.472 |

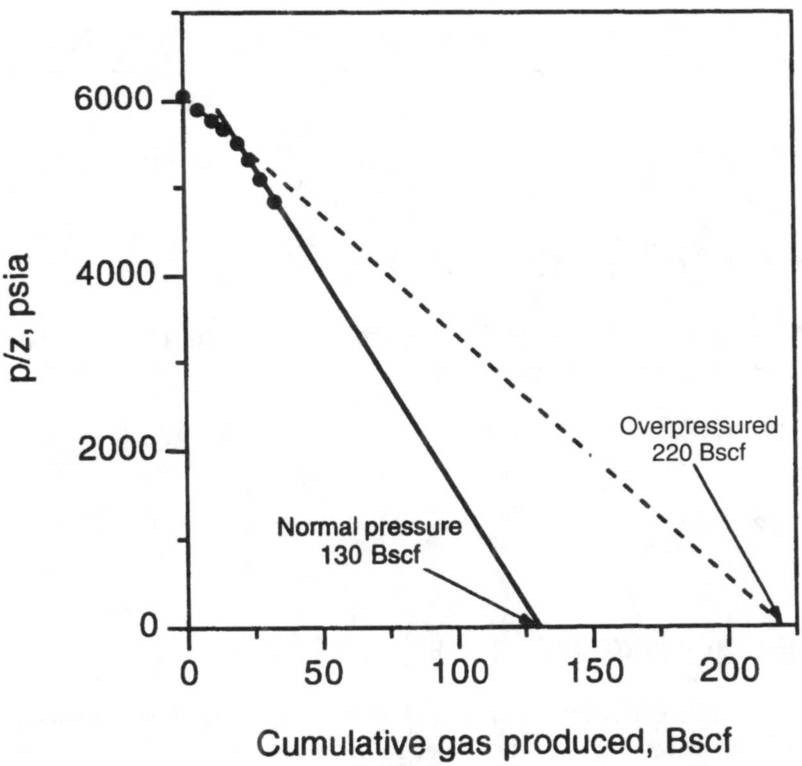


**Fig. 6.1—*p/z* plot for the NS2B reservoir (from Ref. 1).**

*Example Problem 6.1 (continued).* Eq. 6.2 determines the system compressibility at 6,921 psia when the reservoir is assumed to be at normal-pressured conditions.

$$c_s = \frac{(48.7 \times 10^{-6})(0.66) + (3 \times 10^{-6})(0.34) + (21.9 \times 10^{-6})}{0.66}$$

$$= 83.4 \times 10^{-6} \text{ psi}^{-1}.$$

The average-system-compressibility-correction factor is the average of the two previously calculated values.

$$\left(\overline{c_s/c_g}\right) = \frac{\left(c_{efi}/c_{gi}\right) + \left(c_{efn}/c_{gn}\right)}{2}$$

$$= \frac{\left[(64.7.7 \times 10^{-6})/(30 \times 10^{-6})\right] + \left[(83.4 \times 10^{-6})/(48.7 \times 10^{-6})\right]}{2}$$

$$= 1.935,$$

giving

$$G = G_a/\left(\overline{c_s/c_g}\right) = 220 \text{ Bscf}/1.935 = 114 \text{ Bscf}.$$

The other solution method is the change-in-reservoir-volume method. Apply Eq. 6.4 to determine the correction factor.

$$G_{pr} = (302.5 - 281.5)\bigg/\Bigg[(302.5 - 281.5) + \Big(\{281.5(8,921 - 6,921) \times [21.9 + (0.34)3] \times 10^{-6}\}/(1 - 0.34)\Big)\Bigg] = 0.518,$$

giving

$$G = (220 \text{ Bscf})(0.518) = 114 \text{ Bscf}.$$

In the example problem, Hammerlindl[1] demonstrates that either adjustment procedure results in a correct answer. Water-influx effects are assumed to be absent.

A straight line of the later history of the *p/z* plot in Fig. 6.1 extrapolates to OGIP = 130 Bscf. The extrapolated volume is greater than the Hammerlindl adjusted value. However, later performance history indicated that the NS2B reservoir was being affected by water encroachment. The water-influx effect would be to augment reservoir pressure and increase the apparent OGIP estimate.

## 6.3 Ramagost/Farshad Formulation of the Material-Balance Equation

Ramagost and Farshad[2] rewrote the material-balance equation to develop a straightforward calculation procedure amenable to calculator or spreadsheet programs for analyzing the performance history of pressure-depletion, overpressured gas reservoirs. They assumed a constant effective formation compressibility, although the equation is able to account for variable compressibility effects.

**6.3.1 Concept Development**. When the water influx and the *f(p,t)* subordinate energies terms are deleted, the general material-balance equation for an overpressured gas reservoir (Eq. 4.6) reduces to

$$\frac{p}{z}\left[1-\left(\frac{c_w S_{wi}+c_f}{1-S_{wi}}\right)(p_i-p)\right]=\frac{p_i}{z_i}-\frac{p_i G_p}{z_i G}. \qquad (6.5)$$

Eq. 6.5 may be expressed in the form of the equation of a straight line. The components of this straight line are

$$y_{\text{variable}}=\frac{p}{z}\left[1-\left(\frac{c_w S_{wi}+c_f}{1-S_{wi}}\right)(p_i-p)\right] \qquad (6.6)$$

in pounds per square inch, atmospheric,

$$x_{\text{variable}}=G_p \qquad (6.7)$$

in standard cubic feet,

$$y_{\text{intercept}}=p_i/z_i \qquad (6.8)$$

in pounds per square inch, atmospheric,

$$\text{and } m=-p_i/z_i G \qquad (6.9)$$

in psia/scf.

This form of the material-balance equation permits determination of the OGIP when the cumulative-production and formation-, connate-water-, and gas-compressibility values are treated as the input variables. This development is more advanced than Hammerlindl's[1] single-point adjustment methods because the extended performance history is included with the material-balance equation to form a theoretical straight line. The degree of "straightness" allows the quality of the answer to be evaluated.

*Example Problem 6.2.* Duggan[14] and Ramagost and Farshad[2] discussed this example problem of the Anderson L reservoir. **Table 6.2** lists the pressure and production history and reservoir information for the reservoir. The initial 0.85-psi/ft pressure gradient indicates that the reservoir was initially at an overpressured condition. Duggan felt that shale-water influx had a marginal effect in the essentially pressure-depleting reservoir. The volumetric OGIP estimate was 69 Bscf. Ramagost and Farshad compared the normal-pressured *p/z* plot with the more general solution to the material-balance equation for the Anderson L reservoir. The following illustrates the calculation procedure and analyzes the results of the analysis.

*Solution.* Use Eq. 2.3 to calculate the effective formation compressibility. The formation compressibility for the 11,167-ft reservoir was estimated from the Hammerlindl correlation (Fig. 2.11) as $c_f = 19.5\text{x}10^{-6}\ \text{psi}^{-1}$. Fig. 2.10 may also be used.

$$c_e = \frac{S_{wi}c_w + c_f}{1 - S_{wi}} = \frac{0.35(3.4 \times 10^{-6}) + (19.5 \times 10^{-6})}{1 - 0.35}$$

$$= 31.9 \times 10^{-6} + \text{psi}^{-1}.$$

Calculate the *y* plotting variables for both cases. The *p/z* term for the normal-pressure, pressure-depletion case is calculated for each timestep. The *y*-input variable for the Ramagost/Farshad plot is calculated with Eq. 6.6. Table 6.2 presents the results of the calculations for each timestep. An example calculation for the 9,292-psia timestep is

$$p/z = 9,292/1.418 = 6,533 \text{ psia.}$$

$$y_{\text{variable}} = \frac{p}{z}\left[1 - \left(\frac{c_w S_{wi} + c_f}{1 - S_{wi}}\right)(p_i - p)\right] = \frac{9,292}{1.418}$$

$$\times \left\{1 - \frac{(9,507 - 9,292)\left[(3.4 \times 10^{-6})(0.35) + (1.95 \times 10^{-6})\right]}{1 - 0.35}\right\}$$

$$= 6,508 \text{ psia.}$$

*Example Problem 6.2 (continued).* Plot the $y$ variables for both cases vs. cumulative gas production and compare the results. **Fig. 6.2** presents the plot of the results of these calculations. Including formation and connate-water-compressibility as part of the input data in the Ramagost/Farshad solution results in a reasonable straight-line approximation. The $p/z$-plot comparison does not reflect a straight line. The extrapolated OGIP for the Ramagost/Farshad solution is 68 Bscf. Extrapolation of the early-time $p/z$ plot indicates an OGIP value of 108 Bscf. The curved $p/z$ line eventually returns to the volumetric and Ramagost/Farshad extrapolated values. The case study shows how extrapolation of early-time $p/z$ data may result in an optimistic estimate of OGIP.

The Ramagost/Farshad method is useful for analyzing overpressured reservoirs when the pressure- and time-dependent formation-compressibility values are known. However, the problem becomes much more complicated when these parameters are treated as unknowns and when water influx is present. A rearranged version of the general material-balance equation would circumvent this problem.

**TABLE 6.2—ANDERSON L RESERVOIR DATA AND CALCULATION RESULTS (from Ref. 14)**

| Date | $p$ (psia) | $z$ | $G_p$ (Bscf) | $p/z$ | $y_{variable}$* (psia) | $y_{variable}$** ($10^{-5}$ psi$^{-1}$) | $x_{variable}$† (MMscf/psi) |
|---|---|---|---|---|---|---|---|
| January 1966 | 9,507 | 1.440 | 0 | 6,602 | 6,602 | — | — |
| March 1966 | 9,292 | 1.418 | 0.39 | 6,553 | 6,508 | 3.49 | 1.84 |
| July 1966 | 8,970 | 1.387 | 1.64 | 6,467 | 6,357 | 3.88 | 3.12 |
| October 1966 | 8,595 | 1.344 | 3.23 | 6,395 | 6,209 | 3.55 | 3.65 |
| December 1966 | 8,332 | 1.316 | 4.26 | 6,331 | 6,095 | 3.64 | 3.78 |
| January 1967 | 8,009 | 1.282 | 5.50 | 6,247 | 5,949 | 3.79 | 3.88 |
| April 1967 | 7,603 | 1.239 | 7.54 | 6,136 | 5,765 | 3.99 | 4.26 |
| June 1967 | 7,406 | 1.218 | 8.75 | 6,081 | 5,674 | 4.08 | 4.54 |
| August 1967 | 7,002 | 1.176 | 10.51 | 5,954 | 5,479 | 4.34 | 4.65 |
| September 1967 | 6,721 | 1.147 | 11.76 | 5,860 | 5,340 | 4.55 | 4.76 |
| January 1968 | 6,535 | 1.127 | 12.79 | 5,799 | 5,250 | 4.66 | 4.90 |
| March 1968 | 5,764 | 1.048 | 17.26 | 5,500 | 4,845 | 5.35 | 5.54 |
| September 1968 | 4,766 | 0.977 | 22.89 | 4,878 | 4,142 | 7.45 | 6.54 |
| April 1969 | 4,295 | 0.928 | 28.15 | 4,628 | 3,860 | 8.18 | 7.70 |
| October 1969 | 3,750 | 0.891 | 32.57 | 4,209 | 3,438 | 9.88 | 8.87 |
| April 1970 | 3,247 | 0.854 | 36.82 | 3,802 | 3,045 | 11.76 | 10.21 |

$\phi = 24\%$, $S_{wi} = 35\%$, $c_w = 3.4 \times 10^{-6}$ psi$^{-1}$, Datum = 11,167 ft, and $p_i = 9,507$ psia.
*From Eq. 6.6.
**From Eq. 6.12.
†From Eq. 6.11.

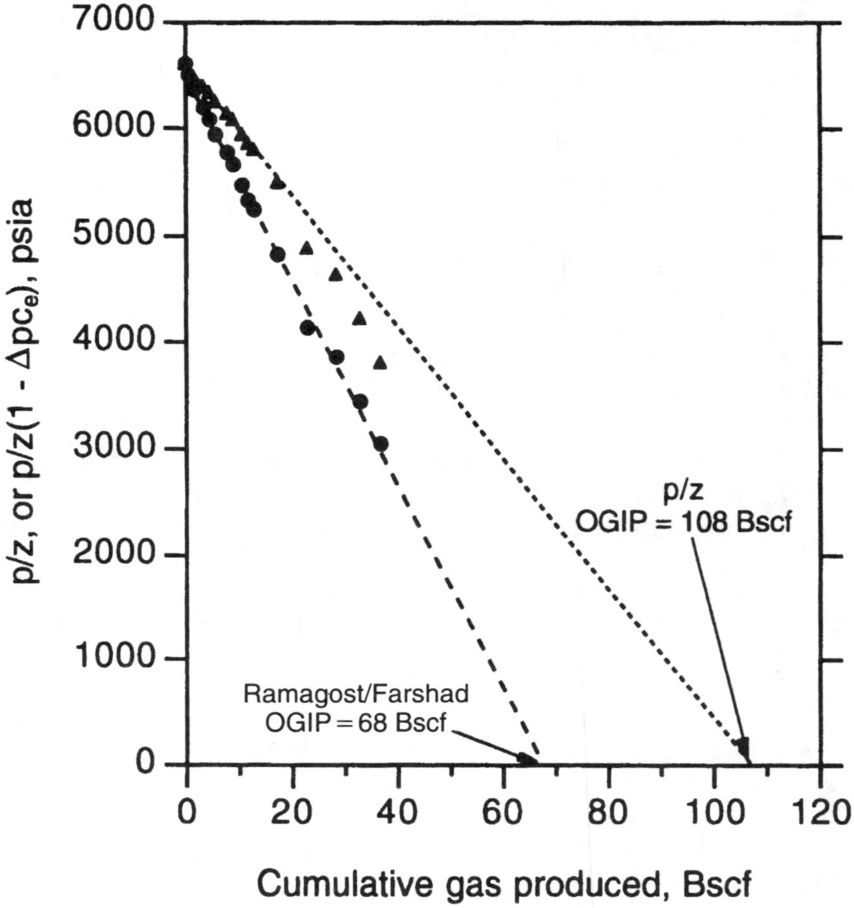


**Fig. 6.2—Comparison of OGIP for the Anderson L reservoir calculated by the *p/z* and Ramagost/Farshad[2] methods.**

### 6.4 Solution-Plot Method

The generalized material-balance equation is rewritten to express an alternative form. The OGIP, formation-compressibility, and water-influx effects are the outcome variables, while production, pressure, and $z$-factor terms are the input terms. This technique was first recognized by Roach.[5] Poston *et al.*[6] and Poston and Chen[15,16] expanded on the utility of the concept by showing that the shape of the data plot also defines the presence or absence of water influx. Additionally, the interpretive method estimates the time frame over which each drive mechanism influences reservoir performance.

**6.4.1 Concept Development.** Eq. 4.6 can be arranged to the following form when the $f(p,t)$ term is assumed irrelevant.

$$\frac{1}{p_i - p}\left(\frac{p_i z}{p z_i} - 1\right) = -\left[\frac{198.4\left(W_e - W_p\right)B_w p_i}{\left(p_i - p\right)G z_i T} + \frac{S_{wi}c_w + c_f}{1 - S_{wi}}\right]$$

$$+ \frac{G_p}{G\left(p_i - p\right)}\left(\frac{p_i z}{z_i p}\right). \quad \text{(6.10)}$$

Applying the equation of a straight line to Eq. 6.10 results in the following plotting components.

$$x_{\text{variable}} = \frac{G_p}{p_i - p}\,\frac{p_i z}{z_i p} \quad \text{(6.11)}$$

in scf/psia

and $$y_{\text{variable}} = \frac{1}{p_i - p}\left(\frac{p_i z}{p z_i} - 1\right) \quad \text{(6.12)}$$

in psia$^{-1}$.

Eq. 6.13 shows that the intercept is a function of the formation-compressibility and water-influx values.

$$y_{\text{intercept}} = -\left[\frac{198.4\left(W_e - W_p\right)B_w p_i}{\left(p_i - p\right)G z_i T} + \frac{S_{wi}c_w + c_f}{1 - S_{wi}}\right] \quad \text{(6.13)}$$

in psia$^{-1}$.

Eq. 6.13 is expressed in a shortened form by applying the definitions for net water influx and effective formation compressibility (Eqs. 4.7 and 2.3, respectively).

$$y_{\text{intercept}} = -\left(c_e + W_{en}\right) \quad \text{(6.14)}$$

in psia$^{-1}$.

The slope, $m$, of the straight-line plot reflects the OGIP.

$$m = 1/G \quad \text{(6.15)}$$

in scf$^{-1}$.

This derivation selects the outcome variables as the unknown reservoir character terms, while the input variables are easily determined from performance and fluid-property rec-

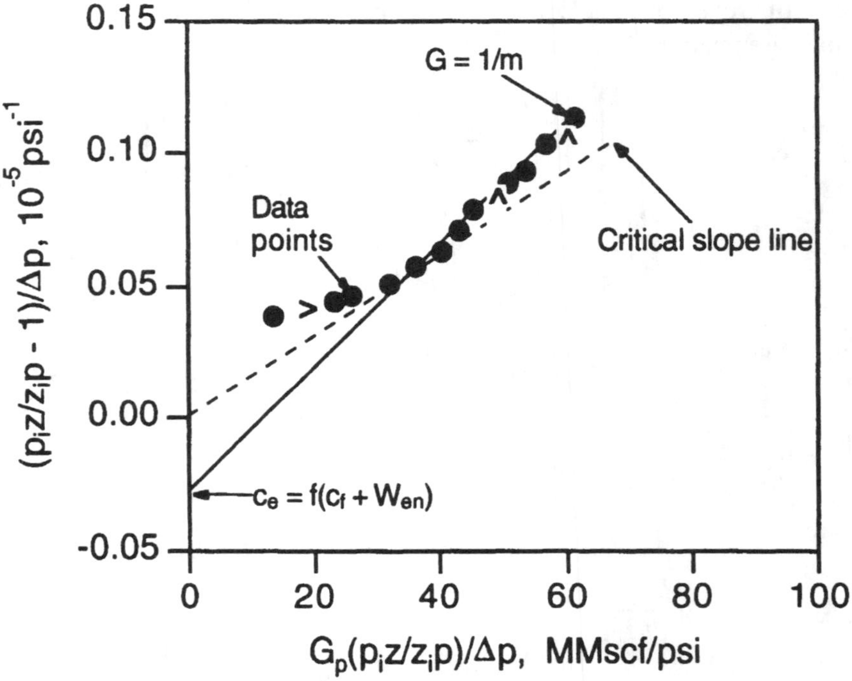


**Fig. 6.3—Example solution plot for the pressure-depleting case (from Ref. 6).**

ords. The OGIP estimate is the inverse of the slope of the straight line and theoretically can be evaluated regardless of the effects of effective formation compressibility and/or water influx on the performance history.

Roach[5] was uncertain of the utility of method because the data plots generally seemed to result in an initial curved line that became straight later in the life of the reservoir. The following discussion develops the meaning of the curved and straight lines and interpretive nature of the solution plot.

**6.4.2 General Solution-Plot Shape Plot. Fig. 6.3** is an example of a solution plot for a typical overpressured, depletion-drive gas reservoir. The curved trend of the early-time data points eventually changes to a diagonally upward straight line. The straight line has been observed in all instances when pressure depletion is the significant drive mechanism. Reservoir simulation studies have shown the straight line reflects steady- or pseudosteady-state depletion conditions, while the curved line represents unsteady-state or transient conditions.[17,18] Transient conditions can reflect changing formation-compressibility or unsteady-state water-influx effects. The dashed line intersecting the origin

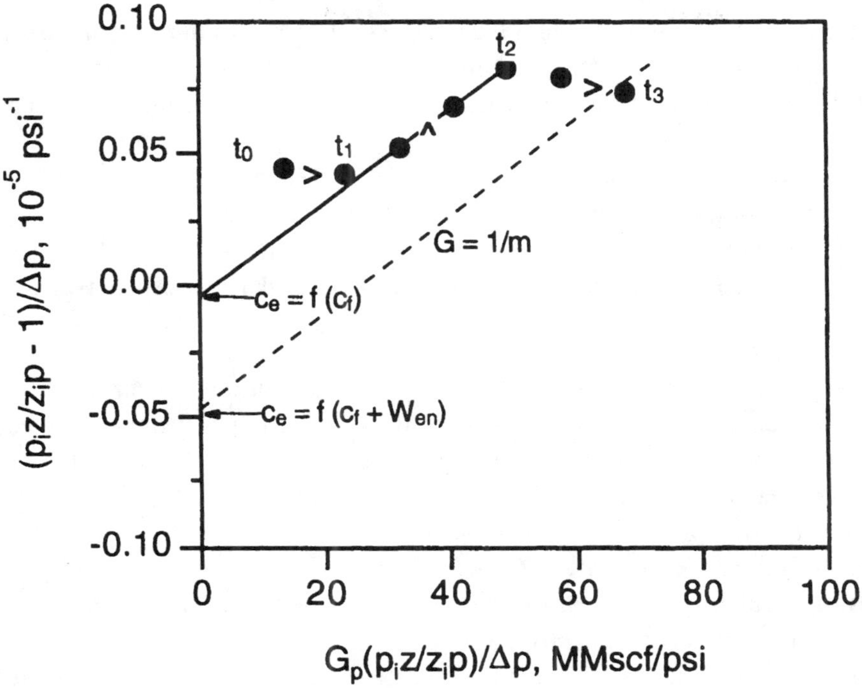


**Fig. 6.4—Effect of water influx on the solution plot; the rightward extension from $t_2$ to $t_3$ represents the influence of water influx (from Ref. 6).**

represents a "critical-slope" line dividing the early-transient from the later steady-state flow regimes. Eq. 6.14 indicates that the intercept value developed from the solution plot must be a negative value. Experience has shown that a positive intersection value above this line usually reflects the very early-time effects of a strong waterdrive. Knowledge of the time and direction of the trend of the solution plot is an equally important consideration. Arrows are included in Fig. 6.3 to indicate the direction of the trend of the data points.

***Effect of Water Influx.*** Simulation studies by Poston and Chen[16] and Hardikar[19] showed that the presence of water influx causes the solution plot to diverge in a rightward direction from the diagonally upward trending straight line. **Fig. 6.4** illustrates a typical divergence. The water-influx effect is significant over the $t_2 \rightarrow t_3$ time period. A transient period spanning the $t_0 \rightarrow t_1$ interval and a steady-state, pressure-depletion period spanning the $t_1 \rightarrow t_2$ interval are also evident in the figure. The volume of water encroachment during the $t_2 \rightarrow t_3$ period is the difference in the $y$-axis-intercept values when lines of equal slope are drawn through the two points of intersection shown as $t_2$ and $t_3$. These extrapolations are shown as the solid and dashed lines in Fig. 6.4. The new intercept value reflects a change in the net water influx because the inverse of the slope of the straight line reflects the unchanging OGIP when effective formation compressibility is assumed to remain constant. The water-influx effect, or rightward trend, of the solution plot may be apparent at very early as well as at late time.

This discussion developed the fundamentals of the mathematical interpretation of the best straight-line fit drawn through the solution plot. Unfortunately, data points often appear to wander and sometimes develop multiple inflections for no obvious reason. Simulation and field case history studies conducted by Poston *et al.*[6] developed the concept of typical curves. These curves interpret the meaning of this sometimes seemingly random behavior.

**6.4.3 Typical Curves.** Interpretation of the shape of the solution-plot curve and the temporal progression of the plotted data points permits evaluation of the reservoir performance history. The presence and/or absence, relative strengths, and time of influence for the pressure-depletion, effective-formation-compressibility, and water-influx drive mechanisms are interpreted from the curves.

***Constant-Compressibility Case—Type 1.*** **Fig. 6.5** shows a typical solution-plot curve for a pressure-depleting, overpressured gas reservoir operating over the $t_1 \rightarrow t_2$ time frame. The value of the $y$ intercept is negative. Simulation studies by Poston and Chen[16] showed that the straight line implies a constant formation-compressibility value in a pressure-depleting reservoir or when water-influx conditions are steady state.

When the reservoir is assumed to be pressure depleting, Eq. 6.13 is reduced to

$$y_{\text{intercept}} = -\frac{S_{wi}c_w + c_f}{1 - S_{wi}} \quad \text{(6.16)}$$

in psia$^{-1}$. To solve for the formation-compressibility term after the intercept has been determined from the solution plot, Eq. 6.16 is rearranged to

$$c_f = -y_{\text{intercept}}(1 - S_{wi}) - (S_{wi}c_w) \quad \text{(6.17)}$$

in psia$^{-1}$. Experience has shown that the calculated formation-compressibility values should range from $6 \times 10^{-6}\,\text{psi}^{-1} < c_f < 20 \times 10^{-6}\,\text{psi}^{-1}$. The presence of a calculated formation compressibility at the high end of the range is probably an indication of a marginally important partial waterdrive.

***Variable-Apparent-Compressibility Case—Type 2.*** An early-time curved line reflects transient conditions. Transient conditions can be attributed to operational and reservoir interactions. The early-time curve shifts if the production rate is increased materially as new completions are added during field development or gas sales are restricted or increased.

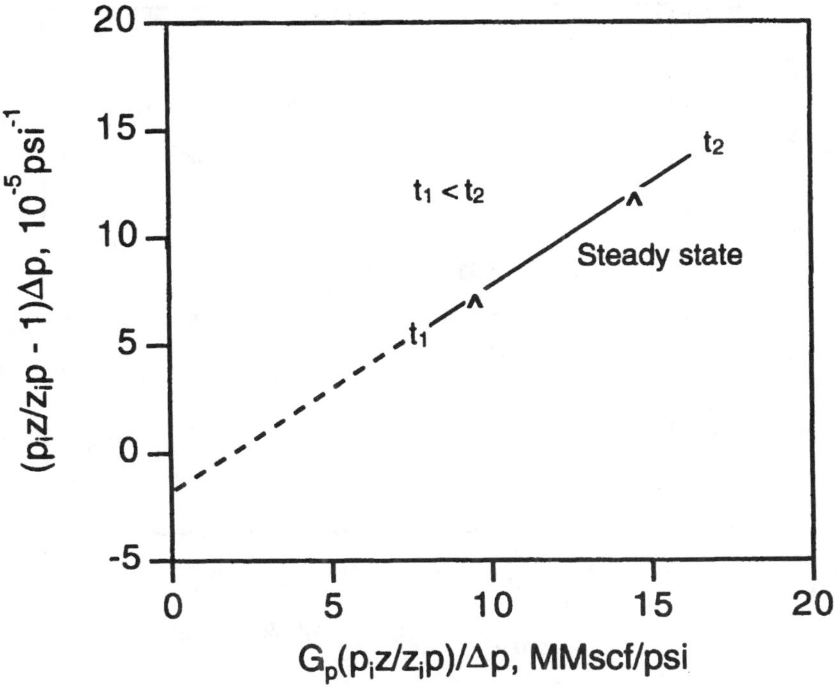


**Fig. 6.5—Type 1 curve for constant-effective-compressibility case where pseudo-steady-state or steady-state conditions always exist (from Ref. 6).**

Pressure loss within a reservoir may cause compaction, influx of shale water, or evolution of gas from the connate or aquifer water to affect the performance history. The effect of shale-water encroachment on the performance history should be at early time, while evolution of solution gas from the connate water would generally be significant at low reservoir pressures. These usually subtle reservoir influences have often been included with the apparent-compressibility term because their influences are usually modest and are extremely difficult to quantify.

Reservoir simulation studies by Hussain[17] indicated that the effect of increasing or decreasing formation compressibility would change the position of the early-time curve in relation to the steady-state straight line. **Fig. 6.6** shows typical curves for the Type 2 case. The arrows illustrate the migration of the data points with respect to time. The curves always trend toward the straight line, even though the initial point may be located on either side of the straight line. The difference between the Type 2a and 2b curves is the origin of the curve at initial time. The following discusses the characteristics of these curve types.

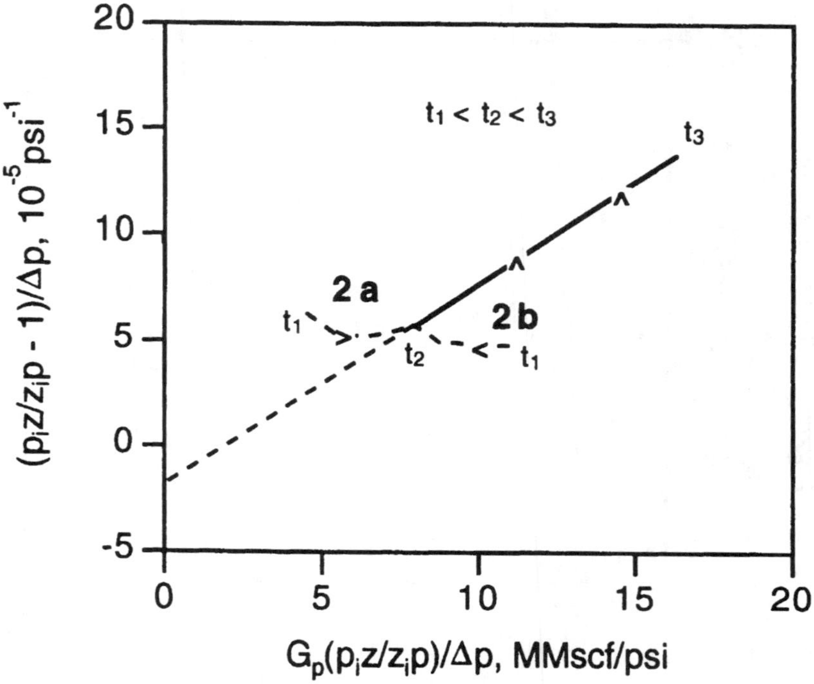


**Fig. 6.6—Type 2 curve for variable-effective-compressibility case trending to steady-state conditions (from Ref. 6).**

*Increasing-Apparent-Compressibility Case—Type 2a.* A typical $p/z$ curve exhibits a concave downward shape at early time. This curve type can be caused (1) by the interaction between the formation, water, and gas compressibility values during the transition from very-high- to moderate-pressure conditions or (2)by inelastic compression of the formation as the pore pressure is reduced by production of reservoir fluids. Shale-water influx could affect the shape of the curve in a similar fashion because this effect would increase and then die out as the shale permeability is reduced by compaction.

*Decreasing-Apparent-Compressibility Case—Type 2b.* A typical $p/z$ curve exhibits a concave downward and then a slight concave upward shape at early time. This curve shape has been noted in two publications. Bruns *et al.*[20] observed this phenomenon when studying gas reservoirs in the presence of very large, permeable aquifers. Wallace's[21] study of waterdrive, overpressured gas reservoirs in South Louisiana indicated several examples of double inflection points in $p/z$ plots.

***Waterdrive Case—Type 3.*** Water influx of increasing, constant, or transient strength may occur. Any rightward deviation away from the straight line is an indication of the

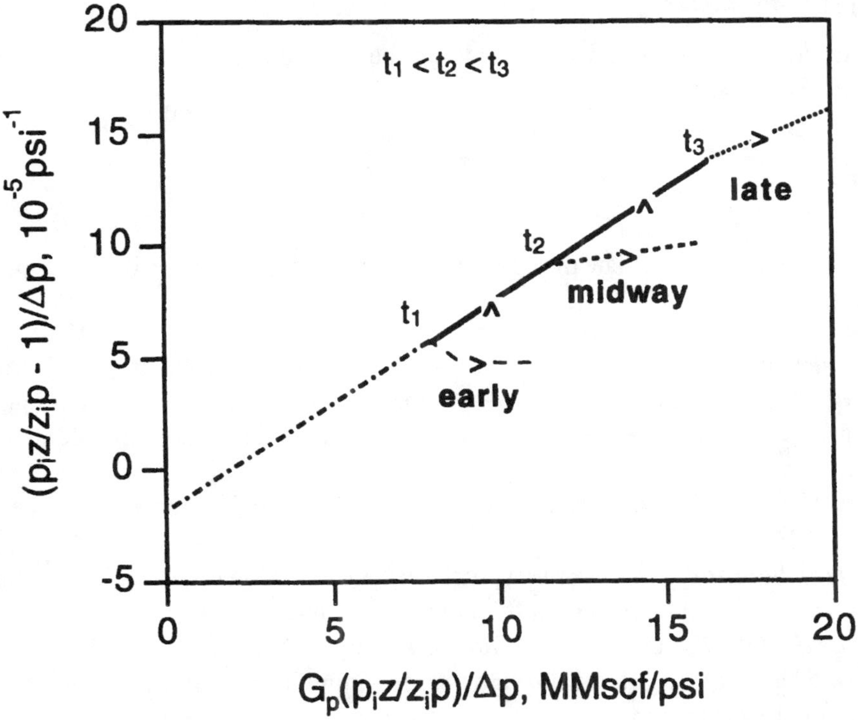


**Fig. 6.7—Type 3 curve for waterdrive case; the rightward extension can occur anywhere along the diagonal line (from Ref. 6).**

effect of water encroachment. **Fig. 6.7** illustrates that water influx may begin at any time during the life of the reservoir and that its presence is reflected by the rightward divergence away from the theoretical diagonally upward trending straight line. A strong, early water-influx effect could cause a rightward extension of the line nearly from the start of production. The effect of water influx on the solution plot may be very subtle; however, a calculated $c_f > 20 \times 10^{-6}$ psi$^{-1}$ usually indicates the probable presence of water encroachment. Arrows in the figure define the direction of migration of the data points.

The following equations reflect the intercept values of the two straight lines in Fig. 6.4. The basket of subordinate energies potential function term is assumed to be negligible, while formation compressibility is assumed to be constant. At $t_2$,

$$y_{\text{intercept},t2} = c_e, \qquad (6.18)$$

and at $t_3$,

$$y_{\text{intercept},t3} = c_e + W_{en}. \quad (6.19)$$

Subtracting Eq. 6.18 from Eq. 6.19 and adding the basic definition for net water influx shown in Eq. 4.7 results in

$$W_e = \frac{(p_i - p)Gz_iT\left[-\left(y_{\text{intercept},t3} - y_{\text{intercept},t2}\right)\right]}{198.4B_wp_i}. \quad (6.20)$$

Eq. 6.20 may be used to calculate the cumulative water influx occurring between any two time periods.

**6.4.4 Analysis Procedure.** The solution plot is constructed from historical-production/reservoir-pressure and pressure/gas-compressibility relationship information. The average initial water saturation and water compressibility are required for calculation of the formation compressibility.

***Construct the Solution Plot.*** Use Eqs. 6.11 and 6.12 to calculate the axes for the solution plot.

***Interpret the Solution Plot.*** Use the typical-curves discussion in Sec. 6.4.3 to interpret the drive mechanisms that appear to affect the performance history.

***Determine the OGIP.*** Draw a straight line beginning at the origin to define the critical slope. Draw a best-fit straight line through the remaining later-time data points trending in a diagonally upward manner. This line represents steady-state conditions. The inverse of this line is the apparent OGIP. Refer to Eq. 6.15. A rightward extension of this second straight line is probably an indication of water influx.

***Calculate Formation Compressibility.*** Include the intercept value in Eq. 6.17 to calculate the apparent formation compressibility. A calculated $c_f > 20 \times 10^{-6}$ psi$^{-1}$ is probably an indication of reservoir pressure enhancement by water influx or from the basket of minor drive mechanisms.

***Estimate Water-Influx Volumes.*** Draw straight lines parallel to the first OGIP interpretation through the water-influx portion of the curve (see Fig. 6.4). The difference in the intercepts of the first OGIP interpretation and subsequent intercepts is a measure of the cumulative-water-influx volume. The following field case histories illustrate the utility of the solution-plot method.

**6.4.5 Reservoir Case History Studies.** ***Anderson L Reservoir.*** The Anderson L reservoir is discussed in Example Problem 6.2. Table 6.2 lists reservoir and performance data. Duggan[14] felt that shale-water influx from a 1,000-ft shale layer immediately overlying the sand influenced the performance history. Aquifer water influx was not present. The analysis procedure is as follows.

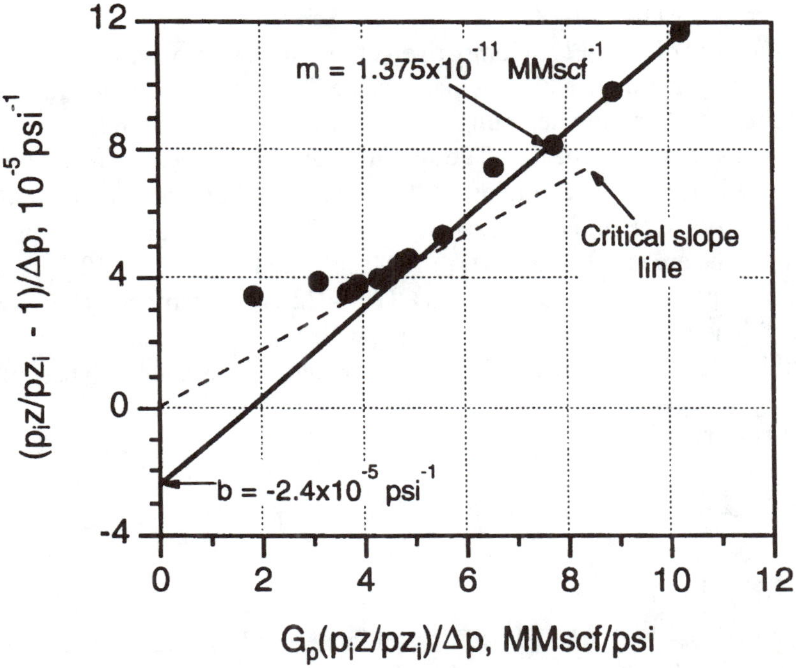


**Fig. 6.8—Solution plot for Anderson L reservoir (from Ref. 15).**

*Calculation of Plotting Variables.* Eqs. 6.11 and 6.12 are used to calculate the *x* and *y* plotting variables for each timestep. Table 6.2 lists the results of the calculations, and **Fig. 6.8** presents the resulting plot. An example calculation for the 8,595-psia timestep illustrates the calculation procedure.

$$x_{\text{variable}} = \frac{G_p}{p_i - p}\,\frac{p_i z}{z_i p}$$

$$= \left(\frac{3.23}{9,507 - 8,595}\right)\left(\frac{9,507}{1.440}\right)\left(\frac{1.344}{8,595}\right) = 3.65 \text{ MMscf/psia}$$

and $$y_{\text{variable}} = \frac{1}{p_i - p}\left(\frac{p_i z}{p z_i} - 1\right)$$

$$= \left(\frac{1}{9,507 - 8,595}\right)\left\{\left[\left(\frac{9,507}{8,595}\right)\left(\frac{1.344}{1.440}\right)\right] - 1\right\}$$

$$= 3.55 \times 10^{-5} \text{ psi}^{-1}.$$

*Interpretation of Solution Plot.* Comparison of the solution plot with the typical curves indicates that the performance history matches the Type 2a curve, which means that the reservoir is probably pressure depleting. There is a well-defined straight line beyond the intersection of the critical slope line. The straight line begins when the reservoir pressure had declined to 7,600 psi. The gradient at this point is still overpressured at 0.68 psi/ft.

*Interpretation of OGIP and $c_f$ Reservoir Values.* The slope, $m$, and intercept, $b$, values for the straight line are $m = 1.375 \times 10^{-11}$ MMscf$^{-1}$ and $b = -2.4 \times 10^{-5}$ psia$^{-1}$. The inverse of the slope of the line results in an OGIP = 72.7 Bscf estimate; Duggan[14] calculated 73 Bscf. The intercept value is combined with Eq. 6.17 to calculate the formation compressibility. Table 6.2 indicates that $S_{wi} = 0.35$ and $c_w = 3.4 \times 10^{-6}$ psi$^{-1}$.

$$c_f = -y_{\text{intercept}}(1 - S_{wi}) - (S_{wi} c_w)$$

$$= \left[-(-2.4 \times 10^{-5})(1 - 0.35)\right] - \left[(0.35)(3.4 \times 10^{-6})\right]$$

$$= 14.4 \times 10^{-6} \text{ psi}^{-1}.$$

The interpreted $c_f = 14.4 \times 10^{-6}$-psi$^{-1}$ value may be a little high. The Hammerlindl[1] correlation in Fig. 2.10 results in a $c_f = 15 \times 10^{-6}$-psi$^{-1}$ estimate. No conclusions can be based on the similarity of these values. The shale-water-influx effect is probably only an early-time event because the shale permeability would be reduced dramatically by compaction. The moderately high value of the interpreted formation-compressibility number probably indicates some marginal effect of water influx on the performance history of the Anderson L reservoir.

***Late-Time Water-Influx Example.*** The initial pressure of the ROB 43-2 reservoir encountered at a 16,488-ft datum was 14,015 psia. Pressure buildup occurred after the reservoir pressure declined to 4,000 psi, when approximately 68% of the OGIP had been produced.[6] The producing rate dramatically declined because casing collapse occurred in many of the wells at the 4,000- to 5,000-psia pressure level. Wallace[21] also observed pressure buildup in waterdrive, overpressured gas reservoirs after an abrupt decline in producing rate. Updip wells generally produced 2 to 4 BWPD, while downdip wells averaged 10 to 20 BWPD. **Table 6.3** lists the reservoir and performance data.

*Construction and Interpretation of Data Plot.* The solution plot in **Fig. 6.9** indicates that steady-state conditions occurred in the middle life of the reservoir when production rates were high. Performance data are not available to study the early-time transient effects. A well-defined rightward trend of the data later in the life of the reservoir is evident. The presence of a Type 3 tail reflects the influence of water influx after the reservoir produc-

**TABLE 6.3—ROB 43-2 RESERVOIR DATA AND RESULTS OF CALCULATIONS**

| Pressure (psia) | $G_p$ (Bscf) | $W_p$ (MMBW) | z factor | x (psia) | y ($10^{-5}$ Bscf/psia) | Intercept ($10^{-5}$ psi$^{-1}$) | $W_e$ (MMBW) |
|---|---|---|---|---|---|---|---|
| 14,015 | 0 | 0 | 1.712 | — | — | | |
| 10,638 | 45.8 | 0.18 | 1.445 | 15.06 | 3.32 | | |
| 5,608 | 114.0 | 0.46 | 1.074 | 21.25 | 6.75 | | |
| 4,968 | 125.6 | 0.50 | 1.033 | 23.65 | 7.77 | | |
| 4,482 | 135.5 | 0.81 | 1.006 | 26.12 | 8.79 | −4.2 | 0 |
| 5,120 | 153.1 | 0.92 | 1.042 | 28.70 | 7.50 | | |
| 5,176 | 153.9 | 0.92 | 1.046 | 28.82 | 7.41 | −6.9 | 21.73 |
| 6,365 | 157.3 | 1.26 | 1.123 | 29.69 | 5.81 | | |
| 6,579 | 157.3 | 1.26 | 1.136 | 29.90 | 5.56 | −9.2 | 33.69 |
| 8,695 | 157.3 | 1.26 | 1.293 | 35.99 | 4.09 | −13.6 | 44.88 |

$\phi = 24\%$, $S_{wi} = 17\%$, $T_R = 280°F = 740°R$, $B_w = 1.05$ RB/STB, and $c_w = 2.1 \times 10^{-6}$ psia$^{-1}$.

tion rate was reduced. Lines intersecting points selected on the water-influx portion of the curve were drawn parallel to the middle-time steady-state region. The similar slopes imply a common OGIP because the effective formation compressibility is assumed to remain constant. Table 6.3 list the values of the $y$-axis intersection.

*Interpretation of Reservoir Parameters.* The slope of the straight line of the steady-state portion of the solution plot is $4.97 \times 10^{-6}$ MMscf$^{-1}$. The inverse of this slope results in an estimate of OGIP = 201 Bscf. Use Eq. 6.17 to calculate the apparent formation compressibility at steady-state conditions.

$$c_f = -y_{\text{intercept}}(1 - S_w) - (S_w c_w)$$

$$= \left[-\left(-4.2 \times 10^{-5}\right)(1 - 0.17)\right] - \left[(0.17)\left(2.1 \times 10^{-6}\right)\right]$$

$$= 34.6 \times 10^{-6} \text{ psi}^{-1}.$$

This moderately high formation-compressibility value probably reflects the presence of a partial waterdrive during the middle life of the reservoir. The initial high producing rate was sufficient to "outrun" most of the effect of the encroaching water during early time. Casing collapse caused dramatic production cutbacks later in the reservoir life. This reduced rate allowed the repressuring effect of the partial waterdrive to become an important influence on the performance history.

*Calculation of Water-Influx Values.* Table 6.3 lists the $y$-intercept values interpreted from Fig. 6.8. These values are entered into Eq. 6.20 to calculate cumulative water influx. An example calculation for the first interval is

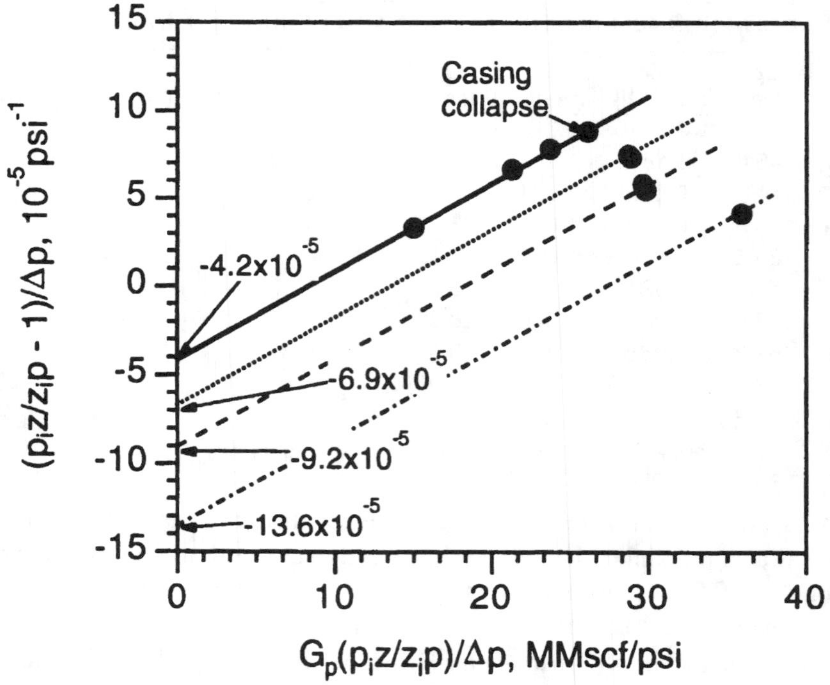


**Fig. 6.9—Solution plot for ROB2 reservoir; the difference between the intersections on the *y* axis represents the water-influx effect (from Ref. 19).**

$$W_e = \frac{(p_i - p)Gz_iT\left[-\left(y_{\text{intercept},t3} - y_{\text{intercept},t2}\right)\right]}{198.4B_wp_i}$$

$$= \frac{(14,015 - 5,176)(201 \times 10^9)(1.712)(740)\left[(6.9 - 4.2) \times 10^{-5}\right]}{198.4(1.05)(14,015)}$$

$$+ 0.92 \times 10^6$$

$$= 21.7 \text{ MMSTB.}$$

Table 6.3 lists the calculated water-influx values.

***Partial Water-Influx Example.*** Harville and Hawkins[13] originally discussed the NS2B reservoir in the North Ossum field, and Hammerlindl[1] used it is as as example problem

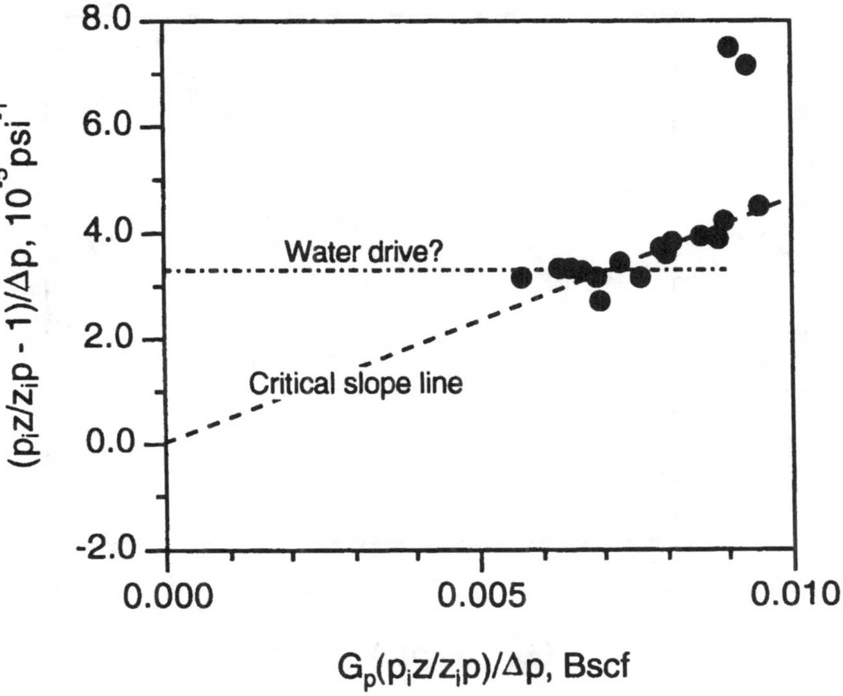


**Fig. 6.10—Solution plot for NS2B reservoir.**

to illustrate the single-point adjustment methods (Example Problem 6.1). Harville[12] lists the reservoir performance history. **Fig. 6.10** shows the solution plot calculated with the NS2B production/pressure information.

The early-time data extend in a nearly horizontal, rightward manner. The critical slope line initiating at the origin extends through essentially all the mid- to late-time data. These two characteristics indicate that water influx is probably a major influence on the performance history. Bernard[7] stated that 5.6 million bbl of water and 60.7 Bscf of gas had been produced from the NS2B reservoir and that there was evidence of a partial waterdrive. Unfortunately, he did not provide the time relationship of the cumulative production values. The solution plot is unable to differentiate between the effects of OGIP and water influx on the performance history of the NS2B reservoir. However, the presence of water influx can be interpreted from the plot.

***Good Waterdrive Reservoir.*** Poston and Chen[15] applied the solution-plot method to study an offshore reservoir with a suspected strong waterdrive. A negative intercept was apparent, but the calculated formation compressibility was $107 \times 10^{-6}$ psi$^{-1}$. The extraor-

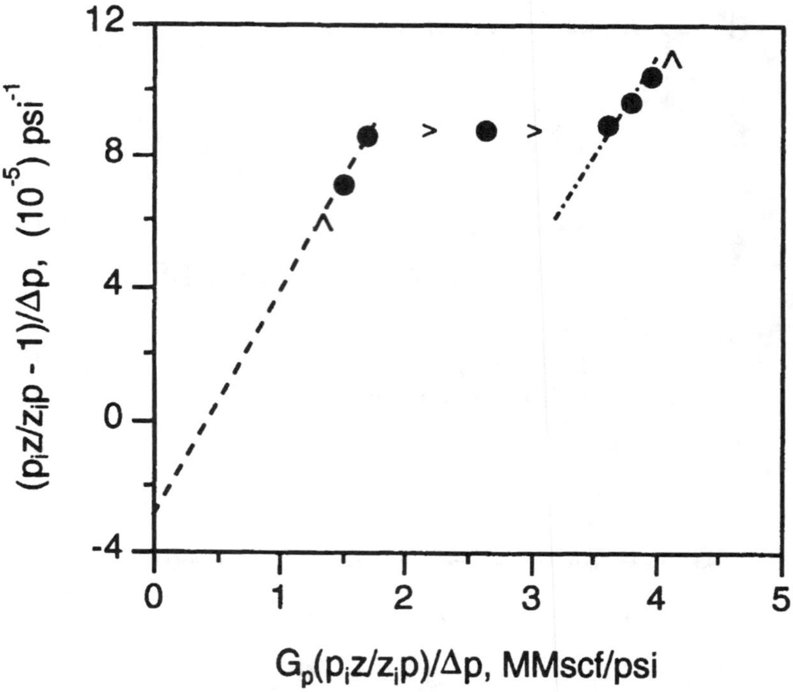


**Fig. 6.11—Solution plot for limited aquifer example [personal communication with D. Barton and B.J. Burke, Chevron U.S.A. Production Co., New Orleans (1981)].**

dinarily high formation-compressibility value indicated the likely presence of a strong, steady-state waterdrive. Both the OGIP and formation-compressibility values should be viewed as suspect, if not completely inaccurate. The solution-plot interpretation method provided an indication of the presence of a strong waterdrive as well as an upper-end estimate of the OGIP.

***Limited Aquifer Example, Offshore Louisiana.*** The DO6, FBA reservoir is an overpressured gas reservoir in a highly faulted region of a Louisiana offshore field.* A downdip water leg was present, and reservoir limits were defined by normal faults. The solution plot in **Fig. 6.11** presents a peculiar three-slope line. The performance history is initially similar to a Type 1 pressure-depletion curve, shifts to a good waterdrive Type 3 curve, and then shifts back to a pressure-depletion Type 1 curve. The essentially similar slopes of the Type 1 lines indicate the influence of a constant original gas volume. Evidently, the aqui-

*Personal communication with D. Barton and B.J. Burke, Chevron U.S.A. Production Co., New Orleans (1981).

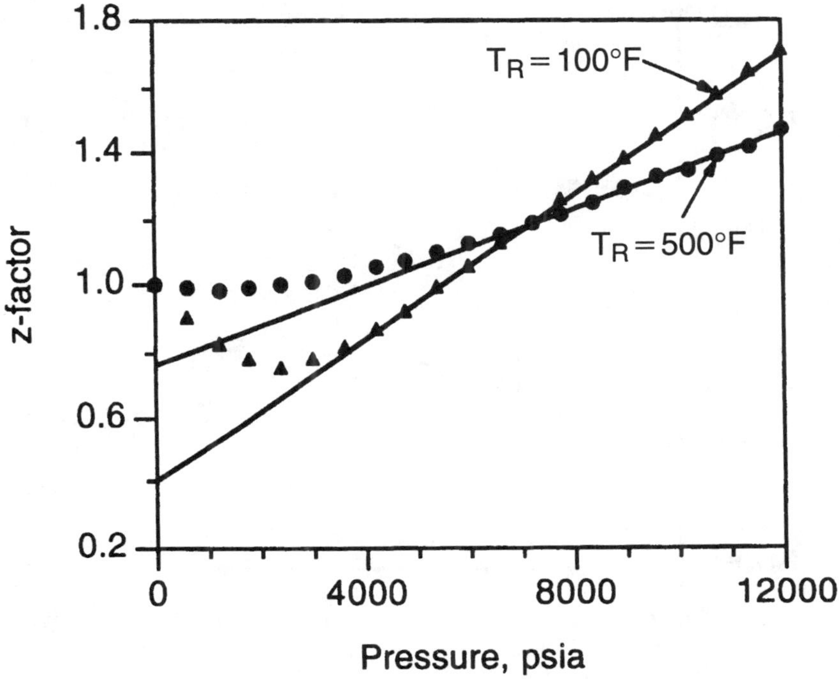


**Fig. 6.12—Illustrating the effect of system temperature on the *p*/*z*-factor relationships for a 0.7-gravity gas (from Ref. 23).**

fer was of limited extent because water-influx energy was influential over only a portion of the life of the reservoir. Structural interpretation verifies the conclusions of a small size gas reservoir with a water leg of limited extent.

### 6.5 Becerra-Arteaga Pressure-Plot Method

Becerra-Arteaga[22] based the pressure-plot method on the observation that a pressure vs. *z*-factor plot forms a straight line at high pressures. **Fig. 6.12** illustrates this effect for a 0.7-gravity gas at 100 and 500°F system temperatures. A straight-line or constant-$(dz/dp)_T$-slope relationship begins at 3,800 psia for the 100°F system temperature and at 6,300 psia for the 500°F system temperature. Each of these temperature-dependent curves may be divided into a high-pressure linear section and a lower-pressure curved section. Becerra-Arteaga developed **Fig. 6.13** to provide a temperature/pressure correlation for estimating the onset of the linear relationship boundaries for 0.6-, 0.7-, and 0.8-gravity gases.

The constant-$(dz/dp)_T$-slope relationship observed at high pressures forms the basis for deriving a material-balance equation for overpressured, depletion-drive gas reservoirs.

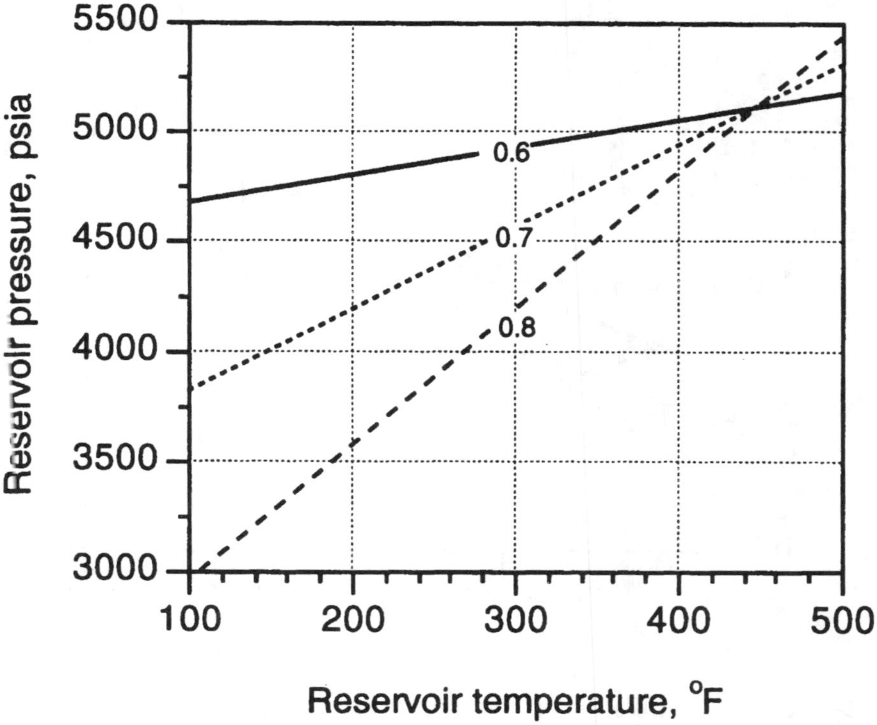


**Fig. 6.13—Correlation for estimating the presence of a straight-line *z* factor for 0.6-, 0.7-, and 0.8-gravity gases (from Ref. 22).**

The temperature- and pressure-dependent $z$ factor is not included in the derivation; therefore, the usual $(p/z)/G_p$ straight-line plot is replaced with a $p/G_p$ straight-line plot. The assumptions forming the principles of the pressure-plot method are that the $(dz/dp)_T$ slope of the $z$-factor plot will remain constant and that water influx does not materially affect reservoir performance.

**6.5.1 Concept Development.** The equation describing the slope of the straight-line relation between pressure and $z$ factor may be written as

$$z = z_i - \frac{\partial z}{\partial p}(p_i - p). \qquad (6.21)$$

Eq. 3.15 developed a relationship between the $z$-factor and gas-compressibility terms.

$$c_{gi} = \frac{1}{p_i} - \frac{1}{z_i}\frac{\partial z}{\partial p}. \qquad (3.15)$$

When Eq. 3.15 is substituted into Eq. 6.21 and the slope terms are set equal to each other, Eq. 6.22 results. The relationship between the gas-deviation and -compressibility terms and pressure changes is

$$z = z_i\left[p_i c_{gi} + \left(\frac{1}{p_i} - c_{gi}\right)p\right]. \quad \text{(6.22)}$$

Eq. 22 expresses the straight-line relationship for the high-pressure and straight-line portion of the Standing and Katz[24] pressure/temperature, $z$-factor chart. Substituting Eq. 6.22 into Eq. 4.10, which is the material-balance equation for an overpressured, depletion-drive gas reservoir results in

$$\frac{G_p}{G} = \left[\frac{(p_i - p)\left(p_i c_{gi} + c_e p\right)}{p_i c_{gi}(p_i - p) + p}\right]. \quad \text{(6.23)}$$

This material-balance equation relates cumulative gas production to OGIP. Note that the $z$-factor term is not included in the equation. All previously developed gas reservoir material-balance relationships include either the gas-formation-volume-factor or the $z$-factor term to account for the changing gas volume measured at reservoir conditions. The equation will be valid as long as the reservoir pressure remains in the straight-line region of the $z$-factor curve.

**Fig. 6.14** illustrates the pressure and $p/z$ plots for an overpressured reservoir depleting to abandonment conditions. Extrapolation of the early-time $p/z$ data results in an incorrect overestimate, $G_o$, of the OGIP, while extrapolation of the early-time pressure data will result in an incorrect underestimate, $G_u$, of the OGIP. The slopes of both curves change to reflect the influence of the increasing gas-compressibility term at moderate to low pressures. Both curves ultimately converge to the true OGIP at atmospheric pressure, $G$. A relationship expressing the underestimate of the early-time extrapolation of the pressure data is

$$G_u = G_p\left(1 + \frac{p}{p_i - p}\right), \quad \text{(6.24)}$$

and a relationship expressing the overestimate of the early-time extrapolation of the $p/z$ data is

$$G_o = G_p\left[1 + \frac{p/z}{(p_i/z_i) - (p/z)}\right]. \quad \text{(6.25)}$$

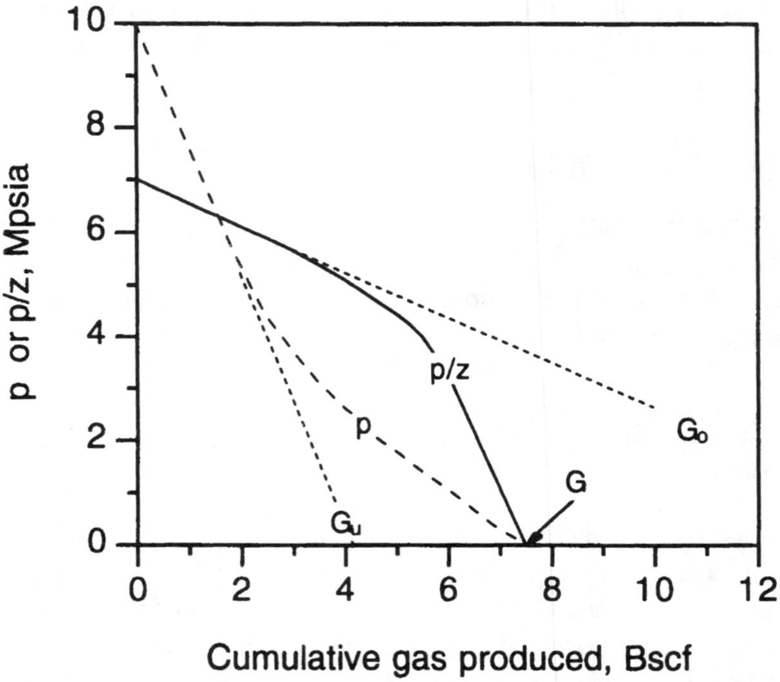


**Fig. 6.14—Comparison of *p/z* and pressure plots for an overpressured, depletion-drive gas reservoir (from Ref. 22).**

The true value of the OGIP estimated at any time must lie in the region between the $G_u$ and $G_o$ boundaries. Therefore, the following relationship can be developed.

$$G = G_p + \gamma\left\{G_p\left[1 + \frac{p/z}{(p_i/z_i) - (p/z)}\right] - G_p\left(1 + \frac{p}{p_i - p}\right)\right\}, \quad \text{(6.26)}$$

where $\gamma$ = percentage of the difference between the two extrapolations. Eq. 6.22 can be rearranged to provide an approximation of *p/z* plot for linear portion of the *z*-factor curve.

$$\frac{p}{z} \approx \frac{p_i}{z_i} - c_{gi}p_i(p_i - p). \quad \text{(6.27)}$$

Eq. 6.28 results when Eq. 6.27 is substituted into Eq. 6.26.

$$\frac{p_i}{z_i}\left(1 - \frac{G_p}{G}\right) = \frac{p_i}{z_i}\left[\frac{ap}{ap + (p_i - p)}\right], \quad \text{(6.28)}$$

where $\alpha$ = average parameter accounting for the linear variation of the $z$ factor lying between the threshold and initial pressure boundaries.

The form of Eq. 28 is similar to the equation of a straight line. The components of the equation are

$$y_{\text{variable}} = \frac{p_i}{z_i}\left[\frac{\alpha p}{\alpha p + (p_i - p)}\right] \quad \text{(6.29)}$$

for the $y$-axis input variable, and

$$x_{\text{variable}} = G_p \quad \text{(6.30)}$$

for the $x$-axis input variable. The slope and intercept values of the equation of the line represent the outcome variables. The slope expression includes the OGIP term and is

$$m = \frac{p_i}{z_i G}, \quad \text{(6.31)}$$

and the $y$-axis intercept is

$$y_{\text{intercept}} = \frac{p_i}{z_i}. \quad \text{(6.32)}$$

The variables included in Eq. 6.28 are pressure, cumulative production, the initial $z$-factor term, and the $\alpha$ term. The first three variables are easily determined from performance information. However, estimating the $\alpha$ term presents a problem.

Becerra-Arteaga[22] conducted an analysis of all available field data to develop an empirical method for finding the $\alpha$ term. The best correlation parameters were found to be the $p_i/z_i$ and the $p_i/D$ terms. **Fig. 6.15** presents the correlation developed from these efforts. Note the effectiveness of the correlation. **Table 6.4** presents the characteristics of the field case histories and references applied to the correlation.

**6.5.2 Analysis Procedure.** Calculate the $p_i/z_i$ and $p_i/D$ terms from the field performance history and reservoir characteristics. Apply these correlation parameters to Fig. 6.15 to determine the proper $\alpha$ term. Two plotting methods can be used to calculate the OGIP.

***Method A.*** Use Eqs. 6.29 and 6.30 to calculate the plotting components for each data point. Plot these values and draw a best straight-line approximation. The intercept on the $x$ axes is the OGIP.

***Method B.*** Eq. 6.28 is rearranged to

$$G = G_p\left[1 + \frac{\alpha p}{(p_i - p)}\right]. \quad \text{(6.33)}$$

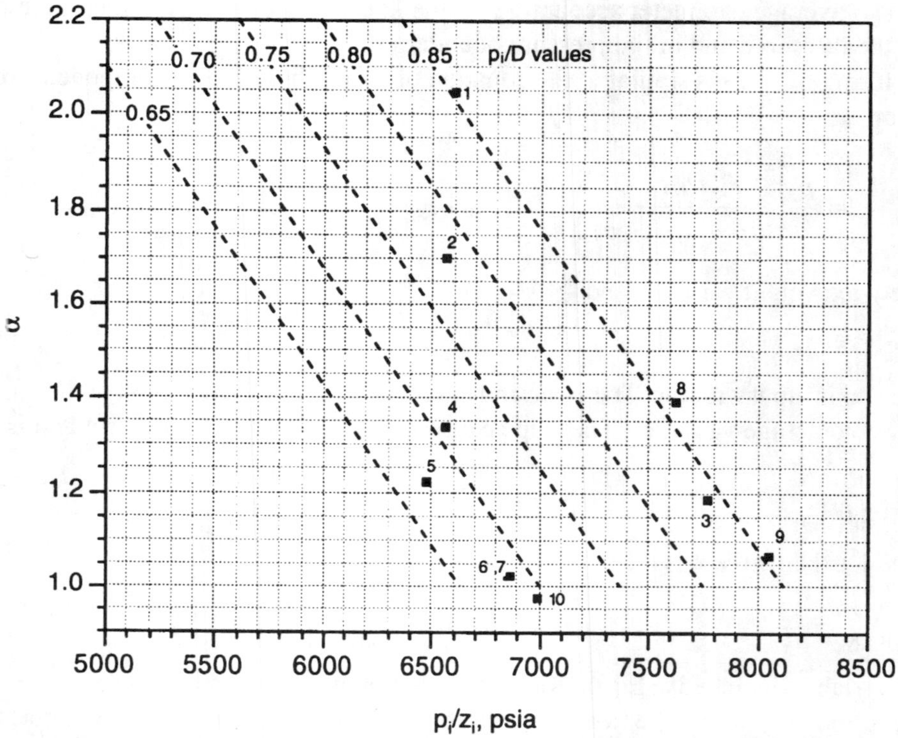


**Fig. 6.15—The $\alpha$ correlation; Table 6.4 shows origins of the numbered correlation points (from Ref. 22).**

**TABLE 6.4—RESERVOIR DATA FOR THE $\alpha$-CORRELATION FIGURE (from Ref. 22)**

| | Field or Reservoir | Reference | $p_i$ (psia) | Depth (ft) | Gradient (psi/ft) | Gravity | $z_i$ | $\alpha$ |
|---|---|---|---|---|---|---|---|---|
| 1 | Anderson L | 14 | 9,507 | 11,167 | 0.85 | 0.94 | 1.44 | 2.05 |
| 2 | Field 117 | * | 8,865 | 11,488 | 0.77 | — | 1.35 | 1.69 |
| 3 | Field 33 | * | 12,000 | 14,242 | 0.84 | 0.60 | 1.54 | 1.22 |
| 4 | Field 38 | *,25 | 8,320 | 11,800 | 0.71 | 0.83 | 1.27 | 1.33 |
| 5 | Field 41 | *,25 | 7,784 | 11,650 | 0.67 | 0.83 | 1.20 | 1.24 |
| 6 | Field 195 | *,25 | 9,260 | 13,700 | 0.68 | 0.60 | 1.35 | 1.00 |
| 7 | Field 268 | *,25 | 9,116 | 13,503 | 0.68 | 0.62 | 1.33 | 1.00 |
| 8 | Cajun | 4 | 11,444 | 13,300 | 0.86 | 0.60 | 1.50 | 1.33 |
| 9 | Rob 43-1 | 6 | 14,015 | 16,488 | 0.85 | — | 1.74 | 1.09 |
| 10 | Field 70 | *,25 | 10,144 | 15,000 | 0.68 | 0.64 | 1.45 | 1.77 |

*Personal communication with W.J. Bernhard, Louisiana State U., Baton Rouge, Louisiana (October 1987).

The calculated OGIP values for each data point are plotted. The best average value is selected for the OGIP estimate.

*Example Problem 6.3.* Table 6.2 lists the reservoir and performance data for the Anderson L reservoir. Find the $a$ correlation term. The correlating terms $p_i/z_i = 6{,}602$ psia and $p_i/D = 0.85$ psia/ft are calculated and applied to the Fig. 6.15 chart to determine the field $a$ correlation value. *Answer*: $a = 2.04$.

*Solution.* Calculate OGIP.

Method 1. Use Eq. 6.29 to calculate the $y$-axis input variable. Table 6.4 lists and **Fig. 6.16** shows the results of these calculations. The intercept of the straight-line extrapolation on the $y$ axis is 72 Bscf. This answer is the same as that determined by Duggan[14] and that obtained by the solution-plot interpretation discussed in Sec. 6.4. The 8,595-psia timestep illustrates the method of calculation for the $y$-axis variable for that timestep. The $x$-axis variable is cumulative gas production.

$$y_{\text{variable}} = \frac{p_i}{z_i}\left[\frac{ap}{ap + (p_i - p)}\right]$$

$$= 6,602\left[\frac{(2.04)(8,595)}{(2.04)(8,595) + (9,507 - 8,595)}\right] = 6,276 \text{ psia.}$$

Method 2. Use Eq. 6.33 to calculate the OGIP for each timestep. **Table 6.5** lists and **Fig. 6.17** illustrates the results of the calculations. There appear to be two levels of average values, with 71 ≦ OGIP, Bscf ≦ 76. The variation of the OGIP values could be caused by minor water influx at late time. An example calculation for the 8,595 psia timestep is

$$G = G_p\left[1 + \frac{ap}{(p_i - p)}\right] = 3.23\left[1 + \frac{(2.04)(8,595)}{(9,507 - 8,595)}\right]$$

$$= 65.3 \text{ Bscf.}$$

In either case, the estimated OGIP approximates that obtained by other methods. The simplicity of the Becerra-Arteaga method is obvious.

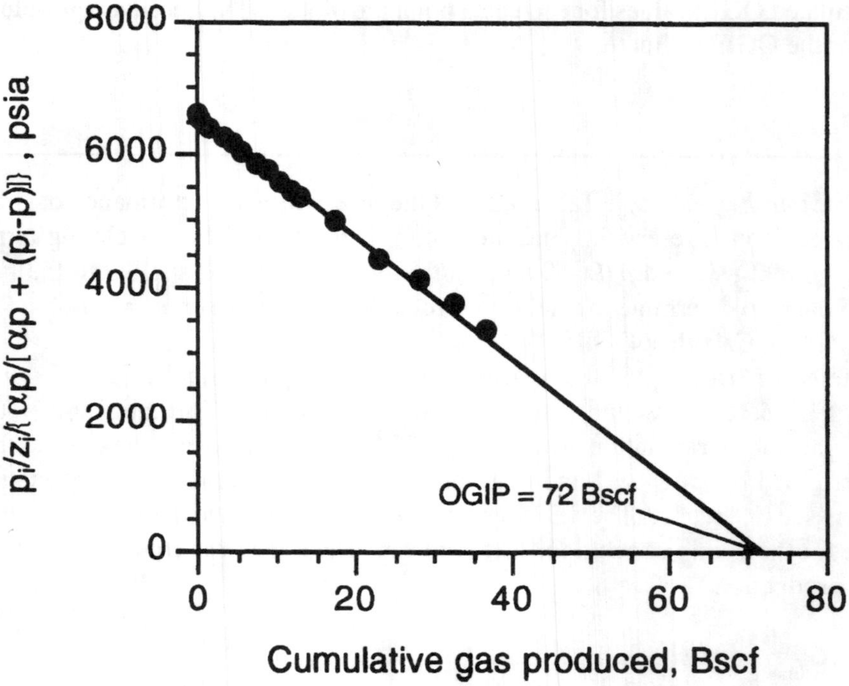


**Fig. 6.16—Method 1 pressure plot solution for Anderson L reservoir (from Ref. 22).**

**Nomenclature**

$A$ = area, $L^2$, $ft^2$
$B_w$ = water formation volume factor, $L^3/L^3$, RB/STB
$c_e$ = effective compressibility, $Lt^2/m$, $psi^{-1}$
$c_f$ = formation compressibility, $Lt^2/m$, $psi^{-1}$
$c_g$ = gas compressibility, $Lt^2/m$, $psi^{-1}$
$c_s$ = system compressibility, $Lt^2/m$, $psi^{-1}$
$c_t$ = compressibility, $Lt^2/m$, $psi^{-1}$
$c_w$ = water compressibility, $Lt^2/m$, $psi^{-1}$
$C$ = well performance coefficient
$C_A$ = Dietz shape factor
$D$ = depth, L, ft
$E$ = error-plot correlation term
$f(p,t)$ = pressure-/time-dependent potential function
$G$ = OGIP, $L^3$, scf

| TABLE 6.5—ANDERSON L RESERVOIR DATA AND RESULTS OF CALCULATIONS | | | |
|---|---|---|---|
| $p$ (psia) | $G_p$ (Bscf) | Method 1—$y_{variable}$ (psia) | Method 2—OGIP (Bscf) |
| 9,507 | 0 | 6,602 | 0 |
| 9,292 | 0.39 | 6,528 | 34.8 |
| 8,970 | 1.64 | 6,414 | 57.5 |
| 8,595 | 3.23 | 6,276 | 65.3 |
| 8,332 | 4.26 | 6,175 | 65.9 |
| 8,009 | 5.50 | 6,048 | 65.5 |
| 7,603 | 7.54 | 5,880 | 69.0 |
| 7,406 | 8.75 | 5,706 | 71.7 |
| 7,002 | 10.51 | 5,617 | 70.4 |
| 6,721 | 11.76 | 5,487 | 69.6 |
| 6,535 | 12.79 | 5,399 | 70.1 |
| 5,764 | 17.26 | 5,008 | 71.5 |
| 4,766 | 22.89 | 4,438 | 69.8 |
| 4,295 | 28.15 | 4,140 | 75.5 |
| 3,750 | 32.57 | 3,767 | 75.9 |
| 3,247 | 36.82 | 3,394 | 75.8 |

$G_a$ = apparent OGIP, $L^3$, scf
$G_o$ = overestimate of OGIP, $L^3$, scf
$G_p$ = cumulative gas produced, $L^3$, scf
$G_r$ = remaining gas volume, $L^3$, scf
$G_u$ = underestimate of OGIP, $L^3$, scf
$m$ = slope of the straight line
$k$ = permeability, $L^2$, md
$p$ = pressure, m/Lt$^2$, psi
$p_a$ = abandonmentpressure, m/Lt$^2$, psi
$p_i$ = initial pressure, m/Lt$^2$, psi
$p_{wf}$ = bottomhole flowing pressure, m/Lt$^2$, psi
$p*$ = iterated pressure, m/Lt$^2$, psi
$\Delta p$ = pressure change, m/Lt$^2$, psi
$q$ = gas flow rate, $L^3$/t, scf/D
$r_d$ = drainage radius, L, ft
$r_e$ = radius to reservoir limit, L, ft
$r_o$ = radius to water contact, L, ft
$r_w$ = wellbore radius, L, ft
$R_D$ = dimensionless aquifer- to reservoir-radius ratio
$R_{s(aq)}$ = solution-gas/-oil ratio of aquifer water, $L^3/L^3$, scf/STB

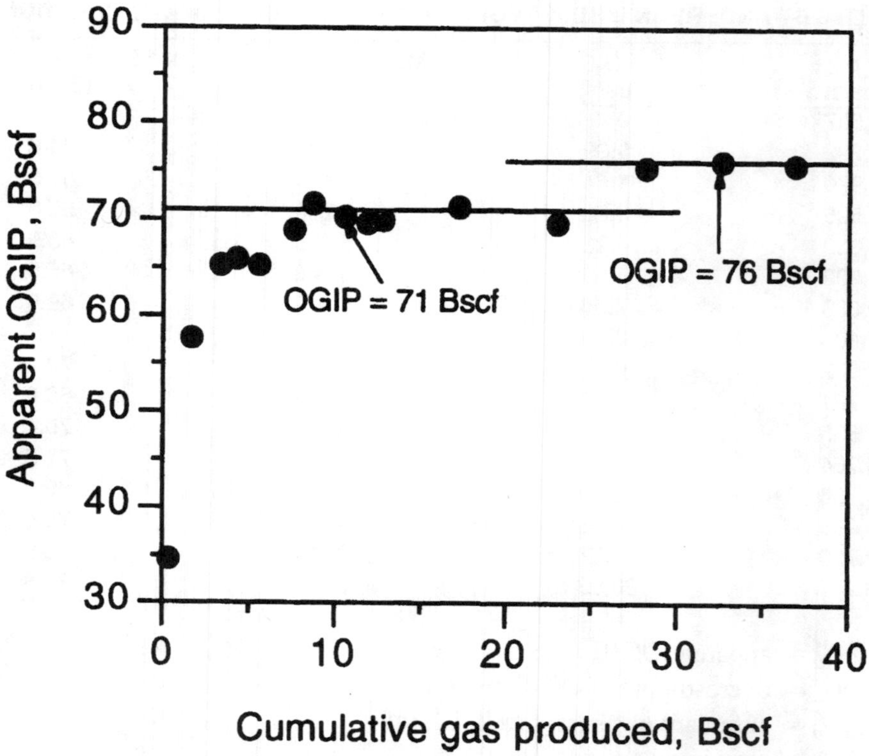


**Fig. 6.17—Method 2 pressure plot solution for Anderson L reservoir.**

$R_{s(cw)}$ = solution-gas/-oil ratio of connate water in shale, $L^3/L^3$, scf/STB
$S_{wi}$ = initial water saturation
$t$ = time, t
$t_s$ = time required to attain static-pressure conditions, t
$T$ = temperature, T, °R
$T_R$ = reservoir temperature, T, °F
$V$ = volume, $L^3$, $cm^3$
$V_p$ = PV, $L^3$, $cm^3$
$V_w$ = water volume, $L^3$, res bbl
$\Delta V_p$ = PV change
$W_e$ = volume of encroaching water, $L^3$, STB
$W_{e(sh)}$ = volume of encroaching shale water, $L^3$, STB
$W_{en}$ = net water influx, $L^3$, STB
$x_{\text{variable}}$ = $x$-axis input variable, $L^2/m\text{-}t^2$, scf/psia

$y_{intercept}$ = $y$-axis intercept, $Lt^2/m$, $psi^{-1}$
$y_{intercept,t2}$ = $y$-axis intercept at time $t_2$, $Lt^2/m$, $psi^{-1}$
$y_{intercept,t3}$ = $y$-axis intercept at time $t_3$, $Lt^2/m$, $psi^{-1}$
$y_{variable}$ = $y$-axis input variable, $Lt^2/m$, $psi^{-1}$
$z$ = gas-deviation factor
$z_i$ = initial single-phase gas-deviation factor
$\alpha$ = ratio of changing $q(\mu c_g)$ terms
$\gamma$ = Euler's constant = 0.5772
$\mu$ = viscosity, m/Lt, cp
$\phi$ = porosity

**Subscripts**

$D$ = dimensionless
$i$ = at initial conditions
$n$ = at later-time conditions

**Superscript**

$\overline{\phantom{x}}$ = average

## References


1. Hammerlindl, D.J.: "Predicting Gas Reserves in Abnormally Pressured Reservoirs," paper SPE 3479 presented at the 1971 SPE Annual Meeting, New Orleans, 3–6 October.
2. Ramagost, B.P. and Farshad, F.F.: "*p/z* Abnormally Pressured Gas Reservoirs," paper SPE 10125 presented at the 1981 SPE Fall Meeting, San Antonio, Texas, 5–7 October.
3. Wang, B. and Teasdale, T.S.: "GASWAT-PC: A Microcomputer Program for Gas Material Balance With Water Influx," paper SPE 16484 presented at the 1987 SPE Petroleum Industry Applications of Microcomputers, Montgomery, Texas, 23–26 June.
4. Begland, T.F. and Whitehead, W.R.: "Depletion Performance of Volumetric High–Pressured Gas Reservoirs," *SPERE* (August 1989) 279; *Trans.*, AIME, **287.**
5. Roach, R.H.: "Analyzing Geopressured Reservoirs—A Material-Balance Technique," paper SPE 9968 available from SPE, Richardson, Texas (1981).
6. Poston, S.W., Chen, H.Y., and Akhtar, M.J.: "Differentiating Between Formation Compressibility and Water-Influx Effects in Overpressured Reservoirs," *SPERE* (August 1994) 183.
7. Bernard, W.J.: "Gulf Coast Geopressured Gas Reservoirs: Drive Mechanism and Performance Prediction," paper SPE 14362 presented at the 1985 SPE Annual Technical Conference and Exhibition, Las Vegas, Nevada, 22–25 September.
8. Ambastha, A.K.: "Analysis of Material-Balance Equations for Gas Reservoirs," paper CIM/SPE 90-36 presented at the 1990 Petroleum Soc. of CIM/SPE International Technical Meeting, Calgary, 10–13 June.
9. Fetkovich, M.J., Reese, D.E., and Whitson, C.H.: "Application of a General Material Balance for High-Pressure Gas Reservoirs," paper SPE 22921 presented at the 1991 SPE Annual Technical Conference and Exhibition, Dallas, 6–9 October.

10. Ambastha, A.K.: "Evaluation of Material Balance Analysis Methods for Volumetric, Abnormally-Pressured Gas Reservoirs," *J. Cdn. Pet. Tech.* (October 1993) **32,** No. 8, 19.
11. Craft, B.C. and Hawkins, M.F.: *Applied Petroleum Reservoir Engineering*, Prentice-Hall Inc., Englewood Cliffs, New Jersey (1959) 487.
12. Harville, D.W.: "Rock Collapse as a Producing Mechanism in Superpressure Reservoirs," MS thesis, Louisiana State U., Baton Rouge, Louisiana (January 1967) 59.
13. Harville, D.W. and Hawkins, M.F.: "Rock Compressibility and Failure as Reservoir Mechanisms in Geopressured Gas Reservoirs," *JPT* (July 1969) 1528.
14. Duggan, J.O.: "The Anderson 'L'—An Abnormally Pressured Gas Reservoir in South Texas," *JPT* (Feb. 1971) 132.
15. Poston, S.W. and Chen, H.Y.: "The Simultaneous Determination of Formation Compressibility and Gas-in-Place in Abnormally Pressured Reservoirs," paper SPE 16227 presented at the 1987 SPE Production Operations Symposium, Oklahoma City, Oklahoma, 8–10 March.
16. Poston, S.W. and Chen, H.Y.: "Case History Studies: Abnormal-Pressured Gas Reservoirs," paper SPE 18857 presented at the 1989 SPE Production Operations Symposium, Oklahoma City, Oklahoma, 13–14 March.
17. Hussain, S.M.: "Analysis of a Geopressured Gas Reservoir Using the Solution Plot Method," MS thesis, Texas A&M U., College Station, Texas (December 1992).
18. Poston, S.W. and Chen, H.Y.: "Using Short Term Pressure Transients to Determine Reserves and Reservoir Character," milestone report, Contract 91–0273, Gas Research Inst., Chicago (April 1992).
19. Hardikar, S.S., "Solution Plot Technique—Analysis of Water Influx in Gas Reservoir Using Simulation Studies," MS thesis, Texas A&M U. (December 1992).
20. Bruns, J.R., Fetkovich, M.J., and Meitzen, V.C.: "The Effect of Water Influx on *p/z*-Cumulative Gas Production Curves," *JPT* (March 1965) 287.
21. Wallace, W.E.: "Water Production From Abnormally Pressured Gas Reservoirs in South Louisiana," *JPT* (August 1969) 969.
22. Becerra-Arteaga , T.: "Analysis of Abnormally Pressured Gas Reservoirs," PhD dissertation, Texas A&M U. (May 1993).
23. Katz, D.L. *et al.*: *Handbook of Natural Gas Engineering* McGraw-Hill Book Co., New York City (1959).
24. Standing, M.B. and Katz, D.L.: "Density of Natural Gases," *Trans.*, AIME (1942) **146,** 140.
25. Prasad, R.K. and Rogers, L.A.: "Superpressured Gas Reservoirs: Case Studies and a Generalized Tank Model," paper SPE 16861 presented at the 1987 SPE Annual Technical Conference and Exhibition, Dallas, 27–30 September.

## SI Metric Conversion Factors

| | | |
|---|---|---|
| bbl | × 1.589 873 | E − 01 = $m^3$ |
| ft | × 3.048* | E − 01 = m |
| $ft^3$ | × 2.831 685 | E − 02 = $m^3$ |
| °F | (°F − 32)/1.8 | = °C |
| psi | × 6.894 757 | E + 00 = kPa |
| $psi^{-1}$ | × 1.450 377 | E − 01 = $kPa^{-1}$ |

*Conversion factor is exact.

# Author Index

# Subject Index